Fluid Mechanics

Mathematical and Mechanical Engineering Set

coordinated by
Abdelkhalak El Hami

Volume 3

Fluid Mechanics

Analytical Methods

Michel Ledoux
Abdelkhalak El Hami

iSTE

WILEY

First published 2017 in Great Britain and the United States by ISTE Ltd and John Wiley & Sons, Inc.

ISTE Ltd
27-37 St George's Road
London SW19 4EU
UK

www.iste.co.uk

John Wiley & Sons, Inc.
111 River Street
Hoboken, NJ 07030
USA

www.wiley.com

Library of Congress Control Number: 2016959733

British Library Cataloguing-in-Publication Data
A CIP record for this book is available from the British Library
ISBN 978-1-84821-951-9

MIX
Paper from responsible sources
FSC® C013604
FSC www.fsc.org

Contents

Preface

Mathematical physics was brought into existence by the development of mechanics. It originated in the study of the planetary motions and of the falling of heavy bodies, which led Newton to formulate the fundamental laws of mechanics, as early as 1687. Even though the mechanics of continuous media, first as solid mechanics, and later as fluid mechanics, is a more recent development, its roots can be found in Isaac Newton's "Philosophiae naturalis principia mathematica" (*Mathematical Principles of Natural Philosophy*), several pages of which are dedicated to the falling streams of liquid.

Applications of fluid mechanics to irrigation problems date back to Antiquity, but the subject gained a key status during the industrial revolution. Energetics was vital to the development of knowledge-demanding, specialized industrial areas such as fluid supply, heat engineering, secondary energy production or propulsion. Either as a carrier of sensible heat or as the core of energy production processes, fluid is ubiquitous in all the high-technology industries of the century: aeronautics, aerospace, automotive, industrial combustion, thermal or hydroelectric power plants, processing industries, national defense, thermal and acoustic environment, etc.

Depending on the target audience, there are various approaches to fluid mechanics. Covering this diversity is what we are striving for in this book.

Whatever the degree of difficulty of the approached subject, it is important for the reader to reflect on it while being fully aware of the laws to be written in one form or another. Various approaches to fluid mechanics are illustrated by examples in this book.

First of all, the student will have the opportunity to handle simplified tools, providing him/her with a convenient first approach of the subject. On the other hand, the practitioner will be provided with elementary dimensioning means.

Other problems may justify or require a more complex approach, involving more significant theoretical knowledge, in particular of calculus. This is once again a point on which students and practitioners who already master these subjects can converge.

A third approach, which is essential for today's physics, especially when dealing with problems that are too complex to be accurately solved by simple calculations, resorts to numerical methods. This book illustrates these remarks.

Problem resolution relies in each chapter on reviews of fundamental notions. These reviews are not exhaustive, and the reader may find it useful to go back to textbooks for knowledge consolidation. Nevertheless, certain proofs referring to important points are resumed. As already mentioned, what matters is that the reader has a good grasp of what he/she writes.

Given that we target wide audiences, the deduction or review of general equations can be found in the appendices, to avoid the book becoming too cumbersome.

The attempt to effectively address audiences with widely varied levels of knowledge, expertise or experience in the field may seem an impossible task.

Drawing on their experience of teaching all these categories of audiences, the authors felt motivated and encouraged to engage in this daring enterprise.

This volume gathers examples of relatively simple approaches to academic problems as well as practical ones. In principle, this work is accessible to all potential readers.

The first chapter recalls the basis of dynamics by focusing on the mechanics of point power. Both the state of fluidity, as well as the main properties of fluids are defined. The problems for writing force, surface and volume when applied to a fluid volume, are discussed. Finally, a strategy for resolving problems in mechanics is approached from a general point of view.

The second chapter covers fluid in equilibrium. The study of incompressible fluid statics under simple forces of gravity or hydrostatics, is completed by that of other forces derived from a potential, such as inertia forces. Compressible fluid statics are also covered.

The third chapter is dedicated to describing flows. The Eulerian vision is favored here. The geometric elements of kinematics are defined. The geometry of flows is established, based on the data of a flow's Eulerian speed. This chapter also provides the opportunity for a first physics principle to be outlined and developed: the principle of continuity.

The first chapter examines the structure of surface forces. This is also where one will find a definition of perfect fluids where the action of viscosity can be overlooked. The fourth chapter is dedicated to processing these flows, in which the Bernoulli theorem is central. Although this theorem is limited by the underlying hypotheses, its strength is observed in how easily one can obtain pertinent orders of magnitude in a large range of phenomena.

When the speed of a fluid varies significantly in a confined space, which is an instance of the barrier between fluids and solids, viscosity becomes a major phenomenon. This is particularly found in pipelines and all components of a hydraulic circuit. In such a situation, one is often only concerned with the loss of mechanical energy that results from fluid friction. This 'head loss' will be calculated in the fifth chapter.

As a general rule, propulsion studies result from a momentum exchange between fluid and a wall. Euler's theorems apply to both perfect flows and viscous fluids and allow one to determine, with a simple knowledge of kinematic fluid passing through boundaries, the resulting moments of a system of forces when applied to a fluid. The sixth chapter will demonstrate how this powerful tool can be applied to determine different types of thrusting.

This work is aimed at students enrolled in engineering schools and technical colleges or in University Bachelors or Masters programs. It is also meant to be useful to the professionals whose activity requires knowledge or mastery of tools related to fluid mechanics.

Michel LEDOUX
Abdelkhalak EL HAMI
November 2016

1

Mechanics and Fluid

1.1. Introduction

The mechanics of fluids is a type of mechanics: it looks at the *movement* of matter when under the influence of *forces*. Matter here is in the "fluid state".

This chapter is approached from the perspective of the foundations of the mechanics of point power. It will also later define what fluid is and which of this matter's main characteristics are useful to know. These characteristics shall then be brought to "life" in later chapters.

1.1.1. *Mechanics: what to remember*

1.1.1.1. *Who is afraid of mechanics?*

For some curious reason, this branch of physics appears frightening to many students, a curse that thermodynamics also shares. Somewhat recoiled from, the mechanical engineer occupies a special place in the academic world. *Some people even wonder whether mechanical engineers are actually physicists who have a strong handle on mathematics, or are in fact mathematicians lost among physicists. These classifications have not been made any simpler by the addition of digital calculations.*

It cannot be stressed enough that the appearance of mechanics gave birth to mathematical physics.

By pairing movement with mathematics, the Neoplanitician, Galileo, created kinematics. And then, with a stroke of genius, although perhaps slightly mythically, Isaac Newton created dynamics by incorporating the fall of an apple and the Moon's trajectory into one vision.

Descartes must not be left out of this Pantheon of emerging physics, for he created momentum, was engaged in heated debates with Newton and Leibnitz on this subject as well as others, and discovered kinetic energy through "life force". Leibnitz and Newton were also the precursors to the differential approach in mechanics.

1.1.1.2. *Principles to remember*

Like a game of chess, the starting rules of mechanics are the simplest. And, like a game of chess, not all paths lead to an easy victory.

a) Remember that a position vector $\vec{r}$ is defined as a vector that links the starting point to another point in space. The coordinates of $\vec{r}$ are evidently the point's three coordinates:

$$\vec{r} = \vec{r}(x, y z) \quad [1.1]$$

By definition, the point's speed is the derivative of the position vector in relation to the time:

$$\vec{V} = \frac{d\vec{r}}{dt} \quad [1.2]$$

which, when passing, accelerates the position vector's second derivative:

$$\vec{\Gamma} = \frac{d\vec{V}}{dt} = \frac{d^2\vec{r}}{dt^2} \quad [1.3]$$

Remember that a vector is derived with regard to a scalar by deriving its components:

$$\vec{r} = [x(t), y(t), z(t)]; \quad \frac{d\vec{r}}{dt} = \left[\frac{dx}{dt}, \frac{dy}{dt}, \frac{dz}{dt}\right] \quad [1.4]$$

b) In 1687, Isaac Newton's *Philosophiae Naturalis Principia Mathematica* outlined three laws, which indeed can be reduced into two:

1) *The principle of inertia*;

2) *Fundamental dynamics law*;

3) *The principle of action and reaction.*

Let us take these three principles further:

Law no. 2. Let us begin with the fundamental dynamics principle, when applied to a constant mass (m) material point:

The acceleration that a body undergoes in an inertial frame of reference is proportional to the resulting forces that it undergoes, and is inversely proportional to its mass.

In modern notation (the notion of the vector was acquired in the 20th Century), this is written as:

$$m\frac{d\vec{V}}{dt} = \vec{F} \qquad [1.5]$$

NOTE: Vectorial notation reminds us that a given speed contains three pieces of information: a direction (instantaneous movement support), a route and an hourly speed. A speed cannot be reduced to the datum of m.s^{-1}. A speed vector not only tells me that my car is traveling at $V = 130\,km.hr^{-1}$ (hourly speed), but it also tells me that I am on a highway between Paris and Rome (direction) and that I am going from Paris to Rome (route). However, I would still need the position vector $\vec{r}$ to tell me where the next exit is.

Therefore, an acceleration is also a vector, and there is no reason why it is not collinear to the speed. Central acceleration in a circular movement is (or should be) known to all secondary school students.

Law no. 1. The principle of inertia was actually discovered by Galileo: *In the absence of an external force, all material points continue in a uniform, straight-lined movement.*

NOTE: This is what Captain Haddock realizes in the "Explorers on the Moon", the illustrated Tintin adventure story by the famous Belgian author, Hergé.

This principle of inertia is in fact a consequence of the fundamental dynamics principle. If the result of forces applied to a material point is zero, then:

$$\vec{F} \equiv \vec{0} \qquad [1.6]$$

and:

$$m\frac{d\vec{V}}{dt}=\vec{0} \quad [1.7]$$

which implies: $\vec{V} \equiv \vec{C}te$ [1.8]

It means a uniform straight-lined movement.

Law no. 3. If the first principle can be reduced to the second, the third principle of action and reaction is independent: *Every body A exerting a force $\vec{F}_{AB}$ on a body B undergoes a force $\vec{F}_{BA}$ of equal intensity, but in the opposite direction, exerted by body B*:

$$\vec{F}_{AB}=-\vec{F}_{BA} \quad [1.9]$$

When solving a problem, to write that every force has an equal and opposite reaction is to write something new with regard to the fundamental dynamics principle.

These principles have been rewritten in various different forms, which lead to equations that are often much more directly applicable. A few of these equations are given in the following sections.

1.1.2. *Momentum theorem*

We can rewrite the fundamental dynamics principle by noting that mass is invariable:

$$m\frac{d\vec{V}}{dt}=\frac{d}{dt}m\vec{V}=\frac{d\vec{p}}{dt} \quad [1.10]$$

A momentum vector has also been introduced:

$$\vec{p}=m\vec{V} \quad [1.11]$$

And the fundamental dynamics principle is also found to be rewritten in terms of momentum:

$$m\frac{d\vec{p}}{dt}=\vec{F} \quad [1.12]$$

In the course of mechanics, it is demonstrated that this equation applies in material points to the center of a group's mass, whether it is continuous or discontinuous and alterable or otherwise. m is therefore replaced by the total mass of the system's points and $\vec{F}$ then represents the resultant of the forces applied to these points. This is what constitutes the center of mass theorem.

NOTE: It goes without saying that we do not intend to write a "digest" here on the course of fluid mechanics.

It would be impossible to attempt to reproduce a complete mechanics course. However, we must insist upon the consequences of these principles which will be directly applied when establishing fluid mechanics theorems. We will build upon the mechanics of point power, and if the reader deems it necessary, they can refer to a dedicated textbook to study system mechanics, which constitutes a more complex domain. Furthermore, in the appendix, we can find a reminder of fluid mechanics equations for a continuous fluid system. This script will be used when demonstrating Euler's first theorem.

We observe that while mass becomes variable with speed, it is this expression that remains valid in particular mechanics. This is also the case for relativist dynamics.

1.1.3. *Kinetic energy theorem*

Forced movement implies work. Here we will give mechanics an energetic dimension. The work of a force $\vec{F}$ when applied to a material point during a time dt provides calculated work from the force and this point's small movement $d\vec{r}$:

$$dW = \vec{F}.d\vec{r} \tag{1.13}$$

$d\vec{r}$ is a small vector, which indicates not only the small distance traveled, but also the carrying line of this movement or direction, and the movement's route. It is linked to speed by:

$$d\vec{r} = \vec{V}\,dt \tag{1.14}$$

Remember the dynamic relation:

$$\vec{F} = m\frac{d\vec{V}}{dt} \tag{1.15}$$

The work is written as:

$$dW = m\frac{d\vec{V}}{dt}.\vec{V}\,dt \qquad [1.16]$$

It can be observed that

$$\frac{dV^2}{dt} = \frac{d\vec{V}.\vec{V}}{dt} = 2.\vec{V}.\frac{d\vec{V}}{dt} \qquad [1.17]$$

Finally, it becomes:

$$dW = \frac{m}{2}\frac{dV^2}{dt}\,dt = \frac{dmV^2}{2} \qquad [1.18]$$

The work performed has helped to increase the quantity $\frac{mV^2}{2}$ carried by the material point. This is how kinetic energy appears:

$$E_C = \frac{mV^2}{2} \qquad [1.19]$$

1.1.4. *Forces deriving from a potential*

In a frame of reference $Oxyz$, where Oz is vertical, the force of gravity $\vec{F}_G$ applied to a mass of $m = 1\ kg$ will have the following components:

$$F_{Gx} = 0 \qquad [1.20.a]$$

$$F_{Gy} = 0 \qquad [1.20.b]$$

$$F_{Gz} = -g \qquad [1.20.c]$$

Furthermore, the operating gradient is defined by associating the vector $\vec{grad}\ f$ with a function $f(x, yz)$ by:

$$\left(\vec{grad}\ f\right)_x = \frac{\partial\phi}{\partial x} \qquad [1.21.a]$$

$$\left(\vec{grad}\, f\right)_y = \frac{\partial \phi}{\partial y} \qquad [1.21.b]$$

$$\left(\vec{grad}\, f\right)_z = \frac{\partial \phi}{\partial z} \qquad [1.21.c]$$

Therefore, $\vec{F}_G$ can be written in the form of a gradient:

$$\vec{F}_G = -\,\vec{grad}\,\phi_G \qquad [1.22]$$

which, by definition, implies the following about the gradient:

$$F_{Gx} = -\frac{\partial \phi_G}{\partial x} = 0 \qquad [1.23.a]$$

$$F_{Gy} = -\frac{\partial \phi_G}{\partial x} = 0 \qquad [1.23.b]$$

$$F_{Gz} = -\frac{\partial \phi_G}{\partial x} = -g \qquad [1.23.c]$$

By identifying:

$$\phi_G = gz + Cte \qquad [1.24]$$

Therefore, it can be said that $\vec{F}_G$ is derived from the potential ϕ_G. It is worth at least being aware of this.

In general terms, it is said that a force $\vec{F}$ is derived from a potential $\phi(x, y, z)$ when

$$\vec{F} = -\,\vec{grad}\,\phi \qquad [1.25]$$

This property is not universal: in particular, friction forces or electromagnetic forces are not derived from a potential.

1.1.5. *Conserving the energy of a material point*

The work performed by a force derived from a potential during a time period of $d\,t$ is written as:

$$dW = \vec{F}.d\,\vec{r} = -\overrightarrow{grad}\,\phi.d\,\vec{r} \qquad [1.26]$$

By developing the scalar product, this can be rewritten in the Cartesian form:

$$dW = -\overrightarrow{grad}\,\phi.d\,\vec{r} = \frac{\partial \phi}{\partial x}dx + \frac{\partial \phi}{\partial y}dy + \frac{\partial \phi}{\partial z}dz \qquad [1.27]$$

The exact total differential is seen to appear ϕ on the time dt, meaning the variation $d\phi$ between the starting point at t and the arrival point at $t + dt$:

$$dW = -\,d\phi \qquad [1.28]$$

By coupling the equations together, we obtain:

$$dW = dE_C = -\,d\phi \qquad [1.29]$$

which can be rewritten as:

$$dE_C + d\phi = 0 \qquad [1.30]$$

Thus, the total energy appears:

$$E_T = E_C + \phi \qquad [1.31]$$

Sum of the kinetic energy and the potential energy, which is conserved when the material point is moving.

NOTE: Remember that when part or all of the forces is or are not derived from a potential, the mechanical energy of the material point is not conserved. The mechanical work of the forces which is not derived from a potential is generally transformed into another form of energy. Thus, friction transforms mechanical energy into thermal energy. This enters into the domain of thermodynamics. The mechanical energy (work) is no longer conserved, but the first principle applies to the two forms of energy: work and heat.

These relations for the material point recalled here have been extended into finite volumes of matter. Curious readers may refer to more elaborate mechanical courses. The aim of this chapter lies in the need for the readers to place themselves within the framework of a basic general culture of mechanics.

All of the notions that have been recalled here will be useful when we begin interpreting Bernoulli's theorem.

1.2. The "fluid state"

The term "fluid state" refers here to the way in which all of the states of matter used to be understood: solid, liquid, gas and plasma, a classification that has been recognized more recently.

In this group, fluid mechanics applies to the last three of these "states".

Solid mechanics deals with alterable and unalterable elastic solids with a blurred boundary and a few creep or pasty rheology problems.

NOTE: It is important not to confuse this expression, which can be traced back to the oldest "fluid state", with the notion of "state in thermodynamics", which relates to a set of thermodynamic variables which we will discuss later.

When approached from the mechanics perspective, this "fluid state" prompts us to:

a) define this state in terms of its nature, its physical qualities and its movements;

b) describe the forces that can be applied to a fluid: what are they and how are they written?

1.2.1. *Fluid properties*

1.2.1.1. *The first property of fluid is its continuity*

Physically, continuity signifies that fluid density, regardless of how small it may be, contains matter. This allows a density to be defined, like the ratio of a small fluid density dm to the small volume $d\omega$ that it occupies:

$$\rho = \frac{dm}{d\omega} \qquad [1.32]$$

For those who like mathematics, we observe that physical continuity connects a notion of continuity for the mass occupying a given volume. This mass $dm(d\,\omega)$ also has a derivative called density. In mathematical terms, the expression is:

$$\rho \text{ exists such that: } \rho = \lim_{d\,Vol \to 0} \frac{dm}{d\,\omega} \qquad [1.33]$$

NOTE: Herein lies a paradox. The mechanical engineer attributes this continuity property to fluid. We know that at the smallest scale of physics, matter is not continuous. Moreover, if there was no fluid discontinuity at the molecular level, we would not be able to determine its essential properties: possible compressibility, existence of pressure and temperature, thermal conduction and matter diffusivity when mixed.

A paradox is merely a poorly asked question. There are at least six or seven orders of magnitude (powers of 10) between the molecular phenomena and the mechanics of a fluids physicist. Admittedly, continuity is just a modeling tool, but it is robust. At the pipeline level, everything happens "as if" the fluid was continuous.

1.2.1.2. *Compressibility*

Density has been defined as a local property. There are many cases where this value of ρ is constant in all fluids. Therefore, it can be said that fluid is incompressible. This will be our definition of incompressibility here. Incompressible is synonymous with $\rho = Cte$.

There are other cases where the density varies from one fluid point to another. Therefore, it can be said that fluid is compressible. This situation is mainly concerned with gas. But the compressibility of liquids may cause certain problems: there is writing on static fluids at the deepest pits of the Pacific Ocean and there are acoustics in liquids (without fluid compressibility, there is no possibility of sound being disseminated).

Determining density, according to its parameters, relates to thermodynamics. All of a fluid state's thermodynamic variables are linked in its equation of state. Subsequently, for a gas, we will need this equation of state. As the equation for so-called perfect gases is the one that is commonly used, we will use it too.

This equation links the three thermodynamic variables: pressure *p*, molar volume M_{mol} or density *p*, absolute temperature *T*, expressed in Kelvin.

This is most often written for a mole (remember that the mole is defined by the number of molecules it contains, namely the Avogadro number $N = 6,022.10^{23}$):

$$pV_{mol} = RT \tag{1.34}$$

Here R is the universal constant of perfect gases, the value of which is $R = 8,3144621 J.mol^{-1}.K^{-1}$. In light of the level of precision of the models, for the examples used in this book, we will choose $R = 8,31 J.mol^{-1}.K^{-1}$. (Some authors use $R = 8,315 J.mol^{-1}.K^{-1}$.)

In mechanics, where the approach is more based on mass, an alternative expression is preferred, which directly uses density as a thermodynamic variable. Therefore, the fluid's molar mass M needs to be brought in.

Noting that

$$\rho = \frac{M}{V_{mol}} \tag{1.35}$$

We obtain the following state equation:

$$pV_{mol} = \frac{pM}{\rho} = RT \tag{1.36}$$

$$pV_{mol} = rT \tag{1.37}$$

with $r = \dfrac{R}{M}$ [1.38]

NOTE: When the ideal-gas law *i* is written under this form, r is no longer a universal constant. It depends on the nature of the fluid.

NOTE: They are called perfect gases because the equation is simple. No gas is intrinsically perfect. This equation is verified by all low-pressure gases. This relation was brought about by the works of Boyle, Mariotte and Charles in earlier times.

"Low-pressure" is a relative expression which may be translated as "any pressure lower than 100 bars in a generous approximation, or 10 bars if one prefers to be pedantic". When looking at compressible fluid problems, we will see that this is a highly acceptable hypothesis.

No fluid is intrinsically incompressible. Contrastingly, a gas can be attributed with an incompressibility property. Incompressible is a synonym of $\rho = Cte$, as is written above. If a flow's conditions are such that ρ varies very little, then $\rho = Cte$ is physically pertinent. Furthermore, we will also see the evolution of pressure strongly coupled with speed. For flows with a relatively weak speed, the pressures vary relatively little and density can easily be deemed a constant. This considerably simplifies the analysis process.

NOTE: For a gas in which the speed scale that establishes a barrier between "incompressible" flows and a "strongly coupled compressible" flow is defined based on the speed of sound in the fluid. Once again, it is in fact the flow that is either "incompressible" or "compressible".

1.2.2. *Forces applied to a fluid*

This section will respond to various questions: how do forces applied to a finite volume of fluid occur? What are these forces and how are they written?

1.2.2.1. *Surface forces, volume forces*

Let us begin with a fluid domain D contained within a closed surface S.

The "exterior" of this domain D will apply two types of forces:

Remote forces which in principle have an application point at all points in the domain D. These are *volume forces*. As a general rule, they are written per mass unit, $\vec{F}_V$.

There are *contact forces* between the external fluid of D and the internal fluid of D. These forces are localized all over the surface S. These are *surface forces*.

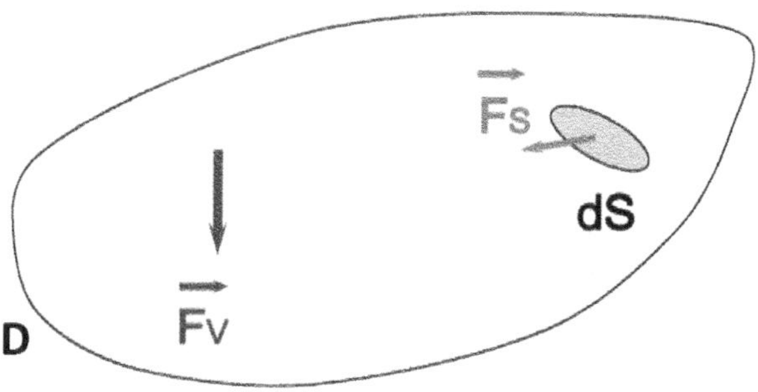

Figure 1.1. *Surface forces, volume forces*

1.2.2.2. *Volume force scripts*

a) As a general rule, volume forces are written per mass unit, $\vec{F}_V$.

These forces are expressed per volume unit as $\rho\,\vec{F}_V$, and the forces applied to a small basic volume $d\omega$ are written as:

$$d\vec{F} = \rho\,\vec{F}_V d\omega \qquad [1.39]$$

b) These volume forces may be "remote" forces, which results in a force field. There are different origins at play here: forces of gravity, which are the most frequent, electrostatic forces and electromagnetic forces. Another type of force will occur when the reference frame, where the problem's equations are written, is no longer inertial (or Galilean).

It is now known that inertia forces appear.

NOTE: Remember that in Newtonian mechanics, an inertial or Galilean frame (they will be used as synonyms here) is a reference frame in a uniform straight-lined movement (meaning in inertial movement, in the sense of Newton's first law) with respect to an absolute frame. In an inertial frame, Newton's second law, along with the absolute frame, applies.

c) The same as for some inertia forces; some remote forces can be derived from a potential.

In this case, a function ϕ_V will be defined as:

$$\vec{F}_V = -\,\vec{grad}\,\phi_V \qquad [1.40]$$

In general, a potential is defined by force nature. If all volume forces are derived from a potential, then the resulting volume forces will also be derived from a potential. This potential function will be the sum at each point in the potential's space of each force. Warning: if just one of the volume forces is not derived from a potential, then the result will not derive from a potential.

Forces of gravity and electrostatic forces are derived from a potential. Electromagnetic forces are not derived from a potential (they are derived from a "vector potential"). Inertia forces, which result from an accelerated translation or a uniform rotation, may be derived from a potential. Examples of this will be given in Chapter 2.

d) Note on calculating inertia forces: Generally, when it comes to a frame with a translation given by a vector $\overrightarrow{OO'}(t)$ and a rotation defined by the rotation vector $\vec{\Omega}$, the inertia force scripts take a complex form. Although we will not be using this expression in all of its complexity here, we will remind the reader of it anyhow.

For a material point of mass m, the inertia force will be:

$$\vec{F}_{inert} = -m\vec{\Gamma}_{rel} \tag{1.41}$$

$\vec{\Gamma}_{rel}$ is calculated based on the different relative movements of the point and the frames.

By appointing $O'x'y',z'$, which is the non-inertial frame where we will solve the problem, where Oxy,z is the Galilean frame against which the frame $O'x'y',z'$ moves and $\vec{r}'$ is the position vector of our material point written in the frame $O'x'y',z'$, we express $\vec{\Gamma}_{rel}$:

$$\vec{\Gamma}_{rel} = \frac{d^2 OO'}{dt^2} + 2\vec{\Omega} \wedge \frac{d\vec{r}'}{dt} + \vec{\Omega} \wedge \left(\vec{\Omega} \wedge \vec{r}'\right) + \frac{d\vec{\Omega}}{dt} \wedge \vec{r}' \tag{1.42}$$

1.2.2.3. *Surface force scripts*

We will discuss two approaches here:

a basic approach, which is sufficient for grasping various problems, and with which all readers of this book should be immediately familiar;

a more complete approach, which we will particularly need for the chapter dedicated to boundary layers, where more complex formulation is required.

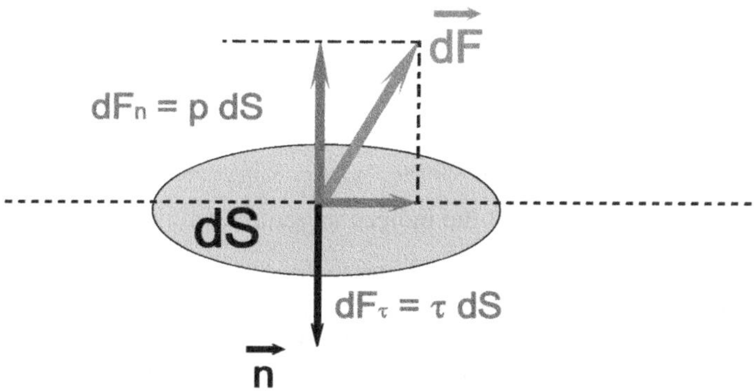

Figure 1.2. *Surface forces: normal forces, tangential forces*

a) Simplified approach

Remember that surface forces are applied from the exterior to the fluid contained within the field, by the fluid immediately in contact with the "internal" fluid at the "*S* level".

That being so, dS is an elementary surface of this surface S and $\vec{n}$ the unitary normal vector to dS.

NOTE: Remember that this unitary vector is carried by the normal force to dS and has a norm equal to 1. Conventionally, this unitary vector is always directed toward the exterior of D, whose purpose is to satisfy the integral vectorial relations.

Two components can generally be distinguished in the surface force $d\vec{F}$ applied to a surface dS:

a normal component carried by $\vec{n}$ and directed toward the interior, $d\vec{F}_n$;

a tangential component $d\vec{F}_\tau$, which is perpendicular to $\vec{n}$, and a tangent to dS.

These two components have an intensity which is proportional to dS. Thus, two finite parameters are defined, the *pressure* p and the *tangential stress* τ such as:

$$dF_n = p\,dS \tag{1.43.a}$$

$$dF_\tau = \tau\,dS \tag{1.43.b}$$

NOTE: The definition of the given pressure must be carefully accepted. In certain instances, the normal component of the volume forces contains terms resulting from the viscosity that is, in principle, reserved for the tangential forces in the previous script. Nevertheless, this will be a pertinent vision for the majority of the applications expanded upon in this book.

b) These forces are a result of the molecular nature of fluid matter

The pressure forces, which are normal to S, result from an exchange of momentum, due to the collision of molecules from the internal and external fluids. This collision is localized to the previously defined S interface. Should a gas interact with a solid wall, such as in a piston for example, which will be familiar to the thermodynamicist, the molecule shocks determine the pressure. Boltzmann modeled this type of mechanics and, in so doing, was able to theoretically establish the law of perfect gases.

It can be demonstrated that the pressure in a fluid point is isotropic. It does not depend on the orientation of the surface dS given by $\vec{n}$.

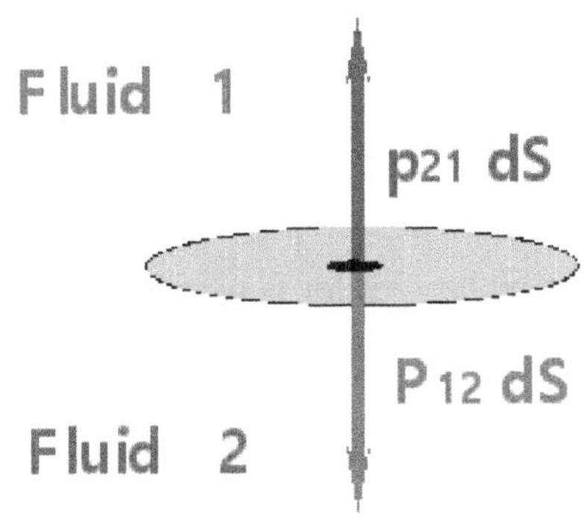

Figure 1.3. *On the interface between the two fluids, the pressure is continuous*

It also demonstrates that the pressure is continuous on the interface between the two fluids 1 and 2. That is, p_{12} is the pressure applied by fluid 1 onto fluid 2, at the level of an elementary surface dS, and p_{21} the pressure applied by fluid 2 onto fluid 1. At the level of dS, the law of action and reaction implies that the action of 1 on 2 is equal and opposite to the action of 2 on 1, in terms of intensity:

$$p_{12}\,dS = p_{21}\,dS; \quad p_{12} = p_{21} \qquad [1.44]$$

Therefore, the pressure is continuous throughout the boundary between fluid 1 and fluid 2.

The tangential forces result from the so-called viscosity phenomenon. This phenomenon was brought to light by the Couette flow experience in its most basic form. This experience, which we will not describe here, allows us to demonstrate that for a flow that is parallel to a flat solid plate, where the speed $u(y)$ varies in a linear way with the distance y to the wall, the tangential stress τ applied by this wall onto the fluid is "most frequently" given by:

$$\tau = \mu \frac{du(y)}{dy} \qquad [1.45]$$

$\frac{du(y)}{dy}$ is very often appointed by the "speed gradient", which is a misnomer. The correct term is shearing.

NOTE: "Speed gradient" is a misnomer because $u(y)$ is a vector component. We will see that the speed can be derived from a potential, but only when there is zero viscosity! In kinematics, an operator $\vec{V}.\vec{grad}\,\vec{V}$ will appear, which is only a means of facilitating the script.

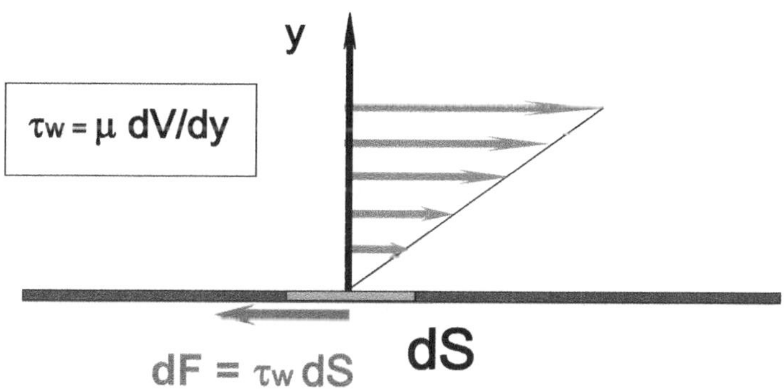

Figure 1.4. *Shearing and stress: Newton's law*

Rheology is the name given to the study of the relationship between stress and shearing.

The linear relationship between stress and shearing [1.45] is fulfilled by the majority of fluid currents, gases or liquids (water, oils, etc.). Therefore, it can be said that fluids fulfill Newton's law or that the fluid is "Newtonian".

μ is therefore defined as dynamic viscosity. It is important to retain this adjective as kinematic viscosity is also defined as the ratio of dynamic viscosity to density. The range of this definition will be discussed further on:

$$\nu = \frac{\mu}{\rho} \qquad [1.46]$$

"Pure" fluids are generally Newtonian. Once the fluid is "charged", meaning when it becomes a solid particle carrier, the rheological behavior becomes more complex and the linear relationship between stress and shearing becomes invalid.

Different models have led to expressions of these elements, some of which are more complex and some less complex (some have been found to be comprised of three lines of equations).

In this instance, we will use a form proposed by Oswald-De Waele:

$$\tau = k\left[\frac{du(y)}{dy}\right]^n \qquad [1.47]$$

where k is a viscosity coefficient.

Examples include non-Newtonian fluids, both of this type and others.

Some fluids are "memorized", such as the so-called Bingham fluids, which can be modeled by:

$$\tau = \tau_0 + k\frac{du(y)}{dy} \qquad [1.48]$$

where τ_0 is a residual stress.

Some fluids have a rheology that varies in time: these are called thixotropic fluids.

NOTE: In practice, the most common non-Newtonian fluids are blood (which contains approximately 45% of solid extract), gels and products (purées, soups) made in the agri-foodstuffs industry. A fluid's memory can also be experienced when we turn a spoon around in a good traditional soup.

c) Scripts developed from tangential stresses. Stresses tensor

This is indeed an oversimplified description. Writing the script correctly requires the stresses tensor σ_{ij} to be defined.

NOTE: The σ_{ij} are actually components of $d\vec{F}_S$ that relate to the three privileged directions, written as i. This indicates that the tension on any surface can be expressed according to the known tensions for the three privileged surfaces dS.

Therefore, the force $d\vec{F}_S$ will be written based on a tension $\vec{\tau}$ such as $d\vec{F}_S = \vec{\tau}\, dS$.

The ith component of $\vec{\tau}$, τ_i will be written as:

$$\tau_i = \sigma_{ij}\, n_j \qquad [1.49]$$

σ_{ij} depends on the fluid's rheology. For a Newtonian fluid, the linearity between stresses and shearing leads to the expression:

$$\sigma_{ij} = \mu \left(\frac{\partial u_i}{\partial x_j} + \frac{\partial u_j}{\partial x_i} \right) + \eta\; div\vec{V} \qquad [1.50]$$

where μ is the dynamic viscosity that we have already seen, which is an essential parameter, and η is the so-called "volume" viscosity, which may be linked to μ.

1.2.2.4. *Surface forces and usual units*

It is recommended that all digital applications are performed in the MKS (meter, kilogram, second) system or the SI system. When using any formula, the parameters must be expressed in SI. We will follow this rule for all of the examples dealt with in this work.

The major parameters in fluid mechanics are always pressure and viscosity. For this reason, beyond legal units, alternative forms have been used to indicate these parameters. We should know all of the different expressions of these parameters in different technical fields.

a) Pressure units.

A pressure is the ratio of a force on a surface. In Newtonian terms, it can be expressed as a meter squared, written as $N.m^{-2}$. A specific unit, the *Pascal,* has been defined and written as Pa.

The pascal represents a very weak pressure. Expressed in water height (see Chapter 2), it is of a $1Pa \approx 0{,}1\,mmCE$ nature. Here, CE signifies a water column. Other units have been defined for the technical dialogue. By using a barometer, we can give a pressure value in liquid "height".

Thus, a *standardized atmosphere* corresponds to a "mercury" pressure of $h = 76\,cm$. Knowing that the mercury density is $\rho = 13600\,kg.m^{-3}$, the pressure in Pascal units will be:

$$1\,atm = \rho g h = 13600 * 9{,}81 * 0{,}76 = 1{,}013.10^5\ Pa \qquad [1.51]$$

The atmosphere emerges at practically $100\ 000\,Pa$. In industrial practice, 1.3% seems to be unchanged, and in this way, the *bar* is defined by:

$$1\,bar = 10^5\ Pa \qquad [1.52]$$

We can incidentally find some old units that are not commonly used nowadays. The reader may find it useful to remind themselves of these by way of reading an older publication.

The CGS (centimeter, gram, second) system has defined some other units. The force unit is the dyne, $1\,dyne = 10^{-5}\,N$. The result is a pressure unit, the dyne per centimeter squared or barye:

$$1\ barye = \frac{10^{-5}\,N}{10^{-4}} = 0{,}1\,Pa \qquad [1.53]$$

We can also cite another old unit, the pièze (pz), which is inherited from the MTS (meter, ton, second) system: $1\,pz = 10^3\ Pa$. We will also notice the hectopièze, which was the pressure unit still used by furnace manufacturers in the middle of the 20th Century. $1\ hpz = 1\ bar$.

Lastly, it should be noted that the millibar appeared in certain domestic barometers.

NOTE: We note that on a good quality barometer, the mercury centimeter scale must lag behind the millibar scale.

b) Viscosity units

There is more than just historical interest in having a knowledge of viscosity units in MKSA (SI) and CGS systems, considering that we can still come across data (particularly in handbooks) that stems from this system. Furthermore, when a viscosity is given, it is rarely indicated whether the viscosity is dynamic or kinematic. Nothing more than the unit is given.

Dynamic viscosity has the following dimension: $ML^{-1}T^{-1}$.

Kinematic viscosity has the following dimension: L^2T^{-1}.

Historically, the names were given in the CGS system.

The dynamic viscosity unit is the *Poise*: $1\ poise = 1\ kg.cm^{-1}.s^{-1}$k.

The kinematic viscosity unit is *stokes*: $1\ stk = 1\ cm^2.s^{-1}$.

The names in MKSA (SI) system were derived from the CGS system.

The dynamic viscosity unit is the *Poiseuille*: $1\ Pl = 1\ kg.m^{-1}.s^{-1}$.

The kinematic viscosity unit is *myriastokes*: $1\ myriastokes = 1 m^2.s^{-1}$.

In principle, myriastokes should appear in an official document. Therefore, it is necessary to know it. Although it is not always known, "Myria" is the significant prefix 10^4. The myriastokes is not very commonly used, as $m^2.s^{-1}$ tends to be the preferred usage.

It is important to know the following conversion:

$$1 poiseuille = 10\, poises; \quad 1 Pl = 10\, ps$$

There is a very important sub-multiple of the poise, that is, the centipoise.

In fact, it is $1 cps = 10^{-3}\ Pl$, which is the order of magnitude of water viscosity.

1.2.2.5. *Perfect fluids. Real fluids*

Studying the aforementioned surface forces at this stage allows us to establish a deep insight into fluid dynamics.

In the simplified presentation, we will break down the normal and tangential components.

There are then three types of situations involved:

In fluid statics, there are no tangential components on the surface forces. Only pressure forces occur.

In fluid dynamics, we find two different cases:

1) The tangential components either do not exist or are negligible. In this instance, they would be perfect fluid dynamics.

2) The tangential components need to be taken into account. In this instance, they must be real fluid dynamics.

Just as accounting for the fluid's compressibility depends on the problem in question, there is no intrinsically perfect or real fluid. We have seen that viscosity forces are linked to the speed "gradients". If these "gradients" are weak, then so are the viscosity forces. *Nevertheless, perfect fluid is associated with the notion of zero viscosity*, even though they are never intrinsically associated!

NOTE: It is not the fluid that is perfect, but the problem. The term "perfection" is just the expression of the physicist's satisfaction with such a simplified problem.

1.3. How to broach a question in fluid mechanics

1.3.1. *The different approaches of fluid mechanics*

Three approaches may be considered for a fluid mechanics problem:

1) *The "table corner" solution,* which is the most basic and the fastest, not necessarily the least formative for the budding or not-quite-so budding physicist.

2) *The complete equation,* the simplification of these equations depending on the proposed problem and the analytical approach. This approach will be adopted when dealing with problems further on.

3) *The digital approach.* This will not be exempt from a prior simplification of the equations in question, whether the calculation means are technically limited or the modeling is vital, as in the case of turbulence.

1.3.2. *Strategies for arriving at a reasoned solution*

1.3.2.1. *The project of this book : give methods*

This book is first and foremost addressed to students, readers who wish to learn, and as such are subject to assessment. We would like to dissuade this type of reader from approaching this work as they would a recipe book.

The examples to be dealt with herein are formalized and complete, in keeping with the "academic" spirit. In principle, they have a solution and all of the elements (data, tables) are provided to avoid wasting time on external research.

NOTE: This method could be criticized within the framework of a certain pedagogy, but research efficiency with regard to time management is also a relevant strategy.

This publication is also aimed at professionals called upon to resolve concrete problems in a professional space, which the student is also destined to do.

In real life, problems may be incomplete or poorly asked. There is often much data that remains to be found.

The following lines attempt to provide an analysis grid that will help readers tackle any problem. Having no intention to revolutionize pedagogical concepts, the methodology presented here results from common sense. It is also implicit to any exceptionally gifted student, to whom there is no need to explain this.

Quite the contrary, once this methodology has been integrated, it will become unconscious and will instead constitute a simple task of reasoning for the young (or not quite so young) fluid mechanical engineer.

1.3.2.2. *What to do when faced with a problem*

Before doing anything else, *know the physical situation*. This reflective activity can be aligned with the plan of this publication.

Then, identify the principles to be written and the laws of the course that are applicable to this situation, including the declensions. This step will enable us to avoid multiple scripts of the same physics law, as well as scripts of inadequate laws.

Recognizing the physical situation is generally a simple operation resulting from common sense.

NOTE: For those using this book, the division of this work by chapter has already been anticipated. Therefore, Any practitioner faced with a problem will be able to refer to a chapter in the following book that is best able to help them.

a) you are faced with an *immobile* fluid: you are dealing with fluid *statics* (Chapter 2).

b) You are given Eulerian characteristics of a flow and you need to find this flow's structure: You are dealing with fluid *kinematics* (Chapter 3).

c) *The fluid flows*.

Is the fluid *compressible or incompressible*: is it a gas or a liquid? Which types of pressure are at play? What are the speeds at play in the flow? If it is compressible, then we must continue with the procedure. Otherwise, we can go directly to the chapter dedicated to compressible flows.

Is it perfect, is it real: what can we say about its viscosity? What do we know about the flow?

If there is zero viscosity and there are no notable speed "gradients", then we have access to the simple solution for perfect fluids dynamics (Chapter 4).

Or, we could be dealing with a pipeline, or something closer to a significant wall. In this instance, we are in the field of real fluids.

So then, what is our objective? To calculate the loss of energy in a pipeline: using well-delineated methods to calculate charge losses will suffice (Chapter 5). Or, is the problem more complex than this? In this case, we need to take a closer look at the flow's structure and understand all types of boundary layers (external, internal and jets).

Perhaps a global approach regarding a system's thrusts (Euler's theorem application) will suffice. In such a case, a chapter dedicated to thrust and propulsion will help us.

Beyond that, in each type of situation, we examine the data we have about the problem and before doing anything else, we ask ourselves what we are looking for.

With regard to writing equations, there is one absolute rule which must always reign: all scripts relate to a principle. We must remain conscious of what we are writing.

The numerous examples that are processed in the following chapters are aimed at helping the reader.

1.4. Conclusion

This introductory chapter has put fluid mechanics back into the general framework of the initial concepts for all kinds of mechanics.

Matter continuously requires a particular approach that led us to specifically formulate the forces applied to matter in this state.

We also wanted to give the reader a larger strategical framework to solve problems, whether they occur in an academic realm or in a more open industrial realm.

We must now return to how these principles are implemented. To do this, in the following chapters, we will need to divide problems according to the analytical framework detailed above. The chapters are divided into the logical segments imposed by this reflection: fluid statics, fluid kinematics, perfect fluid dynamics, real fluid dynamics, broached from various angles; the technical approach to charge loss, the global approach to thrusting, a more analytical approach to flows at borders. After this, we will be able to concentrate on the specifics of compressible flows and then use a digital approach to broach the complexity of flows.

2

Immobile Fluid

2.1. Introduction

The chapter on fluid statics is constituted by the study of immobile fluids. This situation is found every time the system of forces applied to each fluid element is equivalent to a zero force. In what follows from our knowledge of fluid statics, there are two types of problem that mobilize:

1) *Determining the interface's position*, which involves researching the pressure distribution of a fluid group at certain points. This invokes applying the fundamental theorem of fluid statics.

2) *Calculating the thrusts* exerted on a simple or complex surface.

2.1.1. *The fundamental theorem of fluid statics*

The surface forces are reduced to the normal components (as a general rule, tangential components are determined by the velocity field). It is shown that the pressure in one point of a fluid at equilibrium is isotropic (it does not depend on the $\vec{n}$ of the dS surface in question).

There is only one physics principle written on this, which can be outlined thus: *"if a fluid domain is immobile, the resultant forces applied on it are zero"*. Beyond that, this principle can be written in various forms in different hypothesis frameworks.

NOTE.– This fundamental statics principle is actually a particular case of Newton's first principle, or the principle of inertia, when the fluid's velocity is zero.

In the absence of a force, the movement remains unchanged. If the fluid is initially immobile, it continues to have no movement.

a) Hydrostatics

The study of hydrostatics is based on two hypotheses:

fluid is incompressible;

volume forces are reduced to gravity.

A simple demonstration allows us to write a relation that can be generally applied. Let us write the equilibrium of a fluid cylinder with the height of $z_2 - z_1$ in which the bases of area dS are horizontal on the dimensions: z_2 and z_1 on a vertical upward axis. On these dimensions, the pressures are p_2 and p_1, respectively.

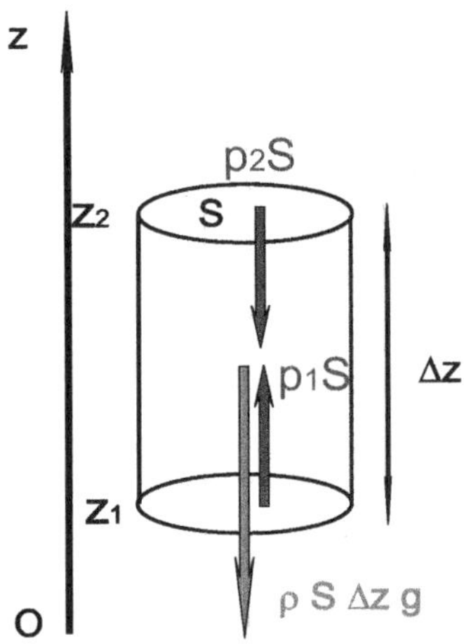

Figure 2.1. *Establishing the fundamental principle of incompressible fluid statics*

The sum of forces applied to this cylinder, projected on a vertical upward axis, is zero:

$$(p_1 - p_2)S - \rho g S(z_2 - z_1) = 0 \quad [2.1]$$

From this, we can easily deduce:

$$p_1 + \rho g z_1 = p_2 + \rho g z_2 \qquad [2.2]$$

Or also

$$p_1 - p_2 = \rho g (z_2 - z_1) \qquad [2.3]$$

Here [2.3] can be translated by using practical reflection: as soon as one "descends" from Δz in a fluid at equilibrium, the pressure increases from $\rho g \Delta z$.

Warning: this is only true if the fluid is incompressible and only under the action of gravity.

b) The case for incompressible fluids and/or any forces derived from a potential

The script for this principle will be more complex, but also more rich in information. We will call upon a differential approach. We will write the equilibrium of a small pavement with dimensions (*dx, dy, dz*). A calculation, which we will not show here, but which uses neither a particular hypothesis on the volume forces, nor on the fluid's compressibility, leads to the following general form:

$$grad\, p + \rho \vec{F}_V = \vec{0} \qquad [2.4]$$

This is not the most commonly used form in problems.

In practice, and as will be the case here, various main assumptions may be used, depending on the complexity of the problem.

Hypothesis H_1: the volume forces deriving from a potential ϕ.

So the fundamental theorem of statics is written as:

$$grad\, p + \rho\, grad\, \phi - \vec{0} \qquad [2.5]$$

Main consequences: the network of isobars (surfaces at $p = Cte$) is confounded with that of equipotentials (surfaces at $\phi = Cte$), and this is not always well retained in conjunction with the network of isochores (surfaces at density $\rho = Cte$). In summary, on a surface $p = Cte$, $\phi = Cte$ and $\rho = Cte$.

An H_2 hypothesis may be added to H_1: fluid is incompressible, which is synonymous with $\rho = Cte$ in the whole fluid area in question.

So, we can write:

$$p + \rho\phi = Cte \quad [2.6]$$

In the particular case where volume forces are reduced to gravity, being *hypothesis H_3*, we are now entering the domain of hydrostatics. Thus, we have an incompressible fluid at equilibrium:

$$p + \rho g z = Cte \quad [2.7]$$

where z is one measured spot altitude on an axis, which must be a vertical upward axis.

There are two very important observations to take into account at this stage:

1) The expression $p + \rho g z$ is only constant as long as ρ is constant while remaining in the same fluid. This quantity varies if we move from one fluid at equilibrium to an adjacent fluid.

2) During the process, it will be seen that it is the local pressure ρ at the interface of two fluids which is continuous.

This is a direct result of the law of action and reaction. At the interface of two fluids 1 and 2, on an elementary surface d, the action of fluid 1 on fluid 2, being p_1dS, is equal and opposed to the action of fluid 2 on fluid 1, being p_2dS. Therefore, the pressures ruling in each field in terms of the interface at dS, p_1 and p_2, respectively, have the same value.

Using these two rules gives us the key to all fluid statics problems.

2.2. Determining the interface position and related questions

2.2.1. *Fluid statics. Incompressible fluids subject to gravity*

EXAMPLE 2.1 (Differential pressure gauge).–

A differential pressure gauge is comprised of a cylindrical tank R, with the diameter D = 5 cm and a cylindrical tube with the length L and diameter $d = 5\,mm$

inclined against the horizontal level of an angle α. The pressures p_1 and p_2, whose difference we want to measure, are applied to A and B. The measurement is performed using a liquid with the following density: $\rho = 10^3\ kg.m^{-3}$.

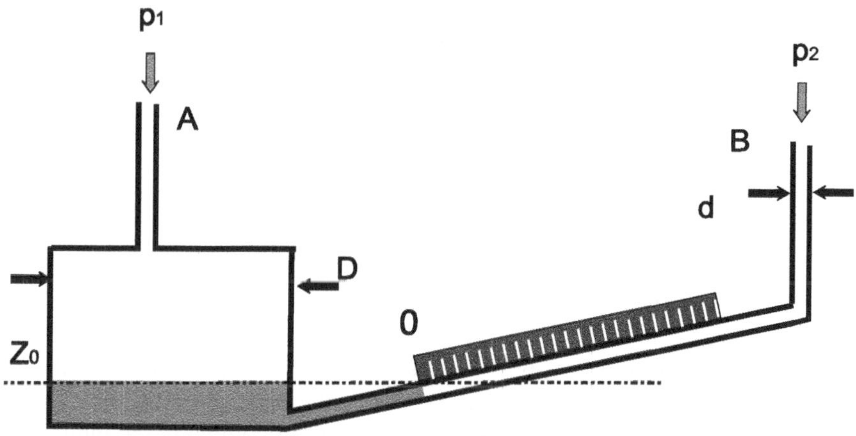

Figure 2.2. *Differential pressure gauge*

First of all, we mark the levels against a horizontal plane with the base of R. The surface level that is against the reference plane and free from liquid, when the atmospheric pressure is applied to A and B, is called z_0.

1) Give the expression of the difference $Z_2 - Z_1$ of the liquid-free surface levels when the pressures p_1 and p_2 are applied to A and B, respectively ($p_1 > p_2$).

2) In practice, d is weak before D. From this, it can be deduced that $Z_0 - Z_1$ is very weak before $Z_2 - Z_0$. Therefore, it gives Z_2.

3) One places a centimeter ruler along the inclined tube. The zero of the ruler is confounded with the level Z_{01}.

Given the expression of the distance measured on the ruler according to the pressures p_1, p_2, of ρ and the angle α, when α is small, what is the value of this device?

In the case where an angle α is equal to 2° and the length of the ruler is equal to 25 cm, what is the maximum pressure difference that can be measured?

Solution:

1) In an incompressible fluid that is only subjected to the forces of gravity, the quantity $p + \rho g Z$ is constant, with Z being one dimension counted on a vertical upward axis.

Therefore:

$$p_1 + \rho g Z_1 = p_2 + \rho g Z_2$$

$$Z_2 - Z_1 = \frac{p_2 - p_1}{\rho g} \qquad [2.8]$$

2) The fluid is incompressible. So the decrease in the volume of the fluid contained in the vessel with a diameter D (due to the passage on the side Z_0 of the free surface at Z_1) must be equal to the increase of fluid volume in the inclined tube (due to the passage of the free surface from Z_0 to Z_2).

NOTE.– The inclined tube has a weak diameter, and the calculation here is confined to considerations regarding the order of magnitude. Therefore, in this calculation, we overlook the effects that inclination has on a free surface against this tube's axis:

$$\pi \frac{D^2}{4}(Z_0 - Z_1) = \pi \frac{d^2}{4}(Z_2 - Z_0)$$

$$\frac{Z_0 - Z_1}{Z_2 - Z_0} = \frac{d^2}{D^2} << 1 \qquad [2.9]$$

In this case:

$$\frac{Z_0 - Z_1}{Z_2 - Z_0} = \frac{d^2}{D^2} << 1$$

$$Z_2 - Z_1 = (Z_2 - Z_0) + (Z_0 - Z_1) \approx Z_2 - Z_0 \qquad [2.10]$$

And $Z_2 - Z_1$ can be directly measured based on the liquid-free surface movement within the inclined tube.

3) The tube's inclination produces an amplifying effect in that the altitude of the liquid-free surface is changed. In fact, a simple trigonometry consideration allows us to link the liquid-free surface movement ΔX along the ruler to the variation of the corresponding dimension ΔZ:

$$\Delta X = \frac{\Delta Z}{\sin \alpha} \qquad [2.11]$$

for $\alpha = 2°$ $\sin\alpha = 3,49.10^{-2}$ and:

$$\Delta X = 28,65\ \Delta Z \qquad [2.12]$$

If $\Delta X = L = 0,25\,m$, then the gap of the maximum dimension is $\Delta Z_{max} = 8,73.10^{-3}\,m$ and the maximum measurable pressure gap Δp_{max} is:

$$\Delta p_{max} = \rho g\, \Delta z_{max} = 85,61\,Pa \qquad [2.13]$$

The sensitivity of the reading is therefore obtained at the cost of a reduction in the measurement range.

This device is still used. Until strain gauge manometers arrived on the market, which are accessed by processing an electronic signal (and computer interfacing), the inclined tube was practically the only means available for measuring weak pressure differentials.

NOTE.– For very weak interface movements, we could also resort to interferometric methods, which need to be implemented very delicately.

EXAMPLE 2.2 (The Hare method for measuring density).–

The upper ends of two vertical tubes, T_1 and T_2, open out into an enclosure E. The lower ends of the tubes are immersed into two different receptacles, each of which contains a liquid. The lower ends T_1 and T_2 are swimming at the fluid density values of ρ_1 and ρ_2, respectively.

The enclosure E is borne at a pressure p_E lower than the atmospheric pressure p_a, which is applied on the liquid-free surfaces 1 and 2.

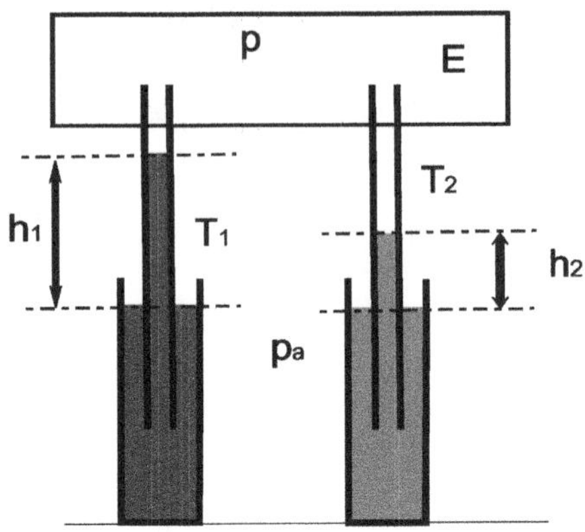

Figure 2.3. *Hare method*

One finds that the free surfaces in tubes 1 and 2 are, respectively, placed at the heights h_1 and h_2 in relation to the free surfaces of the receptacles 1 and 2.

1) Give the expression of ρ_2, knowing ρ_1, h_1 and h_2.

2) Digital application

Liquid 1: water; $\rho_1 = 10^{3 kg.m^{-3}}$.

Liquid 2: benzene.

One measures

$h_1 = 5\,cm$ and $h_2 = 5{,}7\,cm$.

What is the density ρ_2 of benzene?

Solution:

1) The behavior of the liquids in the tubes is independent.

The lower part of each tube is subject to atmospheric pressure and the upper part is subject to the pressure p. The fact that this pressure is the same for both tubes will become valuable when writing the final results.

As always, we mark the altitudes z on a vertical upward axis. Given that we are dealing with fluid statics, in each fluid, the quantity $p+\rho gz$ is constant in each fluid (which, in this case, has its own ρ). Furthermore, the pressures are continuous at the interface of the two fluids.

It is more practical here to use the aforementioned observation:

$p+\rho gz = Cte$ amounts to saying that as soon as one "descends" from z in a liquid at equilibrium, the pressure of ρgz increases.

Therefore, for each tube, we have:

In the tube containing water:

$$p_a = p + \rho_1 g h_1 \qquad [2.14]$$

In the tube containing benzene:

$$p_a = p + \rho_2 g h_2 \qquad [2.15]$$

From this, we can easily deduce the expression of ρ_2:

$$p_a - p = \rho_1 g h_1 = \rho_2 g h_2$$

$$\rho_2 g h_2 = \frac{\rho_1 g h_1}{g h_2}$$

$$\rho_2 = \rho_1 \frac{h_1}{h_2} \qquad [2.16]$$

2) The numerical value of the density is:

$$\rho_2 = \rho_1 \frac{h_1}{h_2} = 1000 \frac{5}{5,7}$$

$$\rho_2 = 877,2\ kg\,m^{-3} \qquad [2.17]$$

EXAMPLE 2.3 (An improvised barometer).–

A physics professor wants to quickly build a barometer for a demonstration. He has no more credit to buy mercury.

He does, however, have some glassware stock left over. Therefore, he undertakes the task of manufacturing a barometer with an oil that has a density of 0.9 $g.cm^{-3}$.

The principle is the same as for all liquid barometers. He fills a sufficiently long glass tube with a diameter $d = 1cm$ and he places it in a tank that already contains 2 l of oil.

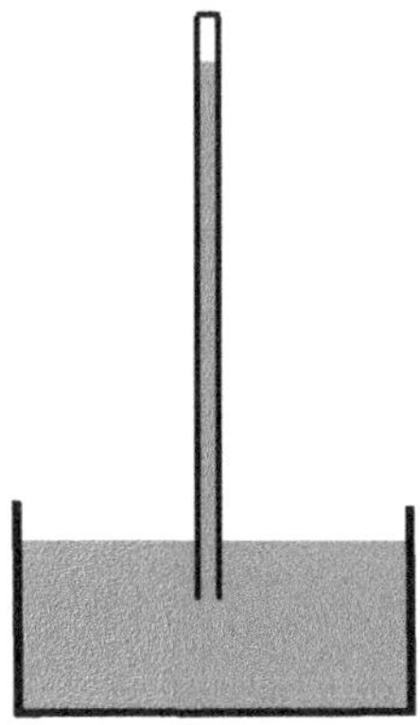

Figure 2.4. *An improvised barometer*

1) How many liters of oil in total must the physicist have in order to measure a pressure of 1 bar?

2) What is the minimum length h that the tube must be in order to measure a maximum atmospheric pressure of 78 cm of mercury if you know that the tube is immersed in the tank by at least 4 cm?

We recall that the density of mercury is $\rho_{HG} = 13600\,kg.m^{-3}$.

Solution:

1) In the tube, we notice that the free surface is at a zero pressure rate.

NOTE.– This constitutes an estimate. Strictly speaking, the pressure would be the oil vapor pressure at the laboratory's temperature.

In terms of the tank, the free surface is at the atmospheric pressure p_a which we wish to measure. Z is the oil height in the tube. In this tube, one descends from Z by increasing the pressure from 0 to p_a. The fundamental theorem of fluid statics allows us to write:

$$p_a - 0 = \rho g Z$$

$$Z = \frac{p_a}{\rho g} \qquad [2.18]$$

If p_a is equal to 1 bar or 10^5 Pa:

$$Z = \frac{10^5}{900\,g}$$

$$Z = 11,33\,m \qquad [2.19]$$

The quantity of oil necessary V_H includes the volume determined by the height Z added to the oil in the tube, or $2.10^{-3}\,m^3$.

The section of the tube emerges at $s = 7,854.10^{-5}\,m^2$:

$$V_H = sZ + 2.10^{-3}$$

$$V_H = 2,89.10^{-3}\,m^3 = 2,89\,liters \qquad [2.20]$$

2) A height of 78 cm of mercury indicates an atmospheric pressure p'_a equal to:

$$p'_a = \rho_{Hg}\,gh = 13600*9,81*0,78$$

$$p'_a = 1,041.10^5\,Pa = 1,041\,bar \qquad [2.21]$$

Using an identical logic to that used in question (1), and with an oil height Z', equal to:

$$Z' = \frac{p'_a}{\rho g}$$

$$Z' = 11,79\,m \qquad [2.22]$$

The minimum length L' of the tube will then be Z' increased by 4 cm:

$$L' = 11,79 + 4.10^{-2} = 11,83\,m \qquad [2.23]$$

EXAMPLE 2.4 (Superimposed liquids. Two-fluid systems).–

The only question to be calculated numerically will only be in question (3).

A large test tube is filled with oil that has a density ρ up until the height h_1. Over the top, one then pours a mercury height of h_2 with a density of ρ_{HG} ($\rho_{HG} > \rho$). The liquids are strictly non-miscible and are adjusted so that the liquid with the density ρ_{HG} remains at equilibrium above the liquid with the density of ρ. (We will also carry out a metastable liquid.) p_a is the atmospheric pressure in the laboratory.

1) What is the pressure p_{int}, at the interface of the two liquids, expressed in pascal?

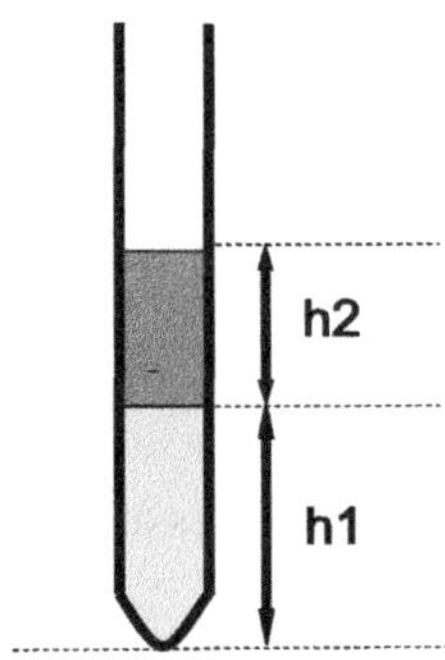

Figure 2.5. *System with two superimposed liquids*

2) One submerges a tube T, which is open at both ends, into the test tube. It is to be done in such a way so that the tube contains no oil, as shown in the diagram.

The lower end of this tube is submerged in the oil. One will find that the oil rises in the tube T up to the height h above the interface of the two liquids.

One would assume that the tube T has a smaller diameter than that of the test tube. And consequently, one can assume that the interface of the two liquids practically does not move when T is entered.

Give the expression of h.

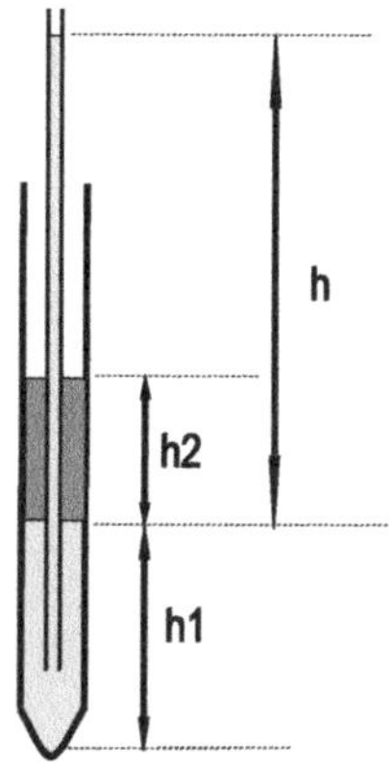

Figure 2.6. *Introducing tube T*

3) Numerical application

We have:

$$p_a = 1\ bar$$
$$h_1 = 10\,cm$$
$$h_2 = 5\,cm$$
$$\rho = 900\,kg.m^{-3}$$
$$\rho_{HG} = 13\,600\,kg.m^{-3}$$

Give the values of p_{int} and h.

Solution:

1) In the mercury, one "descends" from the height h_2 by moving from the atmospheric pressure p_a to the pressure to be determined. One therefore has:

$$p_{\text{int}} - p_a = \rho_{Hg}\, g h_2$$

$$p_{\text{int}} = p_a + \rho_{Hg}\, g h_2 \qquad [2.24]$$

2) The pressure p_{int} calculated in (1) will not be modified by the tube being introduced, which conserves a free surface on the upper part of the mercury.

Furthermore, in the oil, the level constituted by the oil/mercury interface is an isobar level. So, in particular, in the part of this interface located on the inside of the tube, the oil pressure will be p_{int}.

In the oil, between the free surface located on the upper part of the tube and the level of the oil/mercury interface, one descends from h, in order to move from the atmospheric pressure p_a to the pressure p_{int}. One then has:

$$p_{\text{int}} - p_a = \rho g h$$

$$p_{\text{int}} = p_a + \rho g h \quad [2.25]$$

Taking the value of p_{int} into account:

$$p_{\text{int}} = p_a + \rho_{Hg} g h_2$$
$$p_{\text{int}} = p_a + \rho g h$$

$$h = h_2 \frac{\rho_{HG}}{\rho} \quad [2.26]$$

3) The numerical values of p_{int} and h emerge at:

$$p_{\text{int}} = p_a + \rho_{Hg} g h_2 = 10^5 + 13600 * 9,81. * 0,05$$

$$p_{\text{int}} = 1,0667.10^5 \; Pa \quad [2.27]$$

$$h = h_2 \frac{\rho_{HG}}{\rho} = 0,05 \frac{13600}{900}$$

$$h = 0,7556 \, m = 75,56 \, cm \quad [2.28]$$

EXAMPLE 2.5 (Superimposed liquids. Three-fluid systems).–

Let us now consider a tank into which we have poured water with a density of $\rho_W = 1000 \, kg.m^{-3}$ up to a height $h_1 = 100 \, cm$, and an oil with a density $\rho_0 = 900 \, kg.m^{-3}$ up to a height $h_2 = 150 \; cm$.

The atmospheric pressure is $p_a = 1 \; bar$.

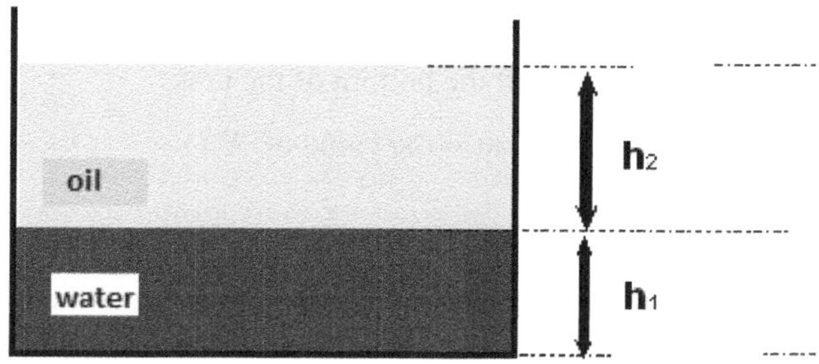

Figure 2.7. *The tank with two liquids*

1) What is the pressure p_0 at the bottom of the tank?

One introduces a vertical tube into the liquid so that the lower end is immersed in the water and 50 cm from the bottom. Then, one adds into the tube an unknown height h_3 of mercury with a density of $\rho_{HG} = 136000\,kg.m^{-3}$. As the tank has a large surface, one will consider that h_1 and h_2 remain strictly constant throughout the whole problem.

2) Give the expression according to h_3 of the pressure p_{HG0} in the tube at the interface between the mercury and the oil.

3) Give the expression according to h_3 of the pressure p_{OW} in the tube at the interface between the oil and the water.

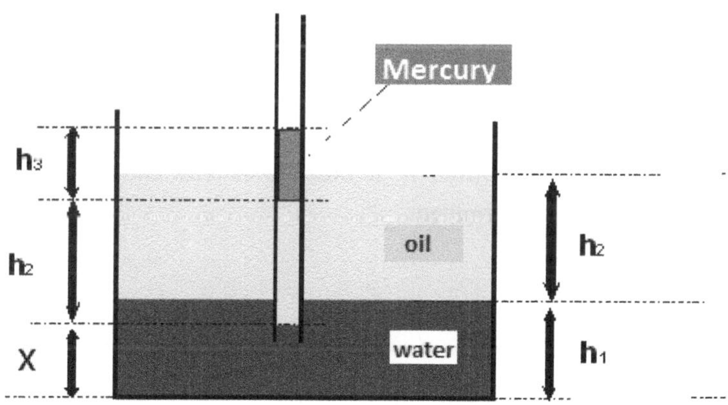

Figure 2.8. *The three-fluid system*

4) Give the expression according to h_3 of the altitude X of the interface between the oil and the water against the bottom of the tank.

5) If one produces $h_3 = 4 cm$, what is the value of X?

Solution:

1) Let us call the pressure at the interface between the oil and the water p_{int}.

In the oil, as soon as one "descends from h_2", the pressure increases from the atmospheric pressure p_a to the pressure p_{int}:

$$p_{\text{int}} - p_a = \rho_o g h_2 \qquad [2.29]$$

In the water, as soon as one "descends from h_1", the pressure increases from the pressure p_{int} to the pressure p_{0t} to be determined:

$$p_0 - p_{\text{int}} = \rho_w g h_1 \qquad [2.30]$$

In the end:

$$p_0 = p_a + \rho_w g h_1 + \rho_o g h_2 \qquad [2.31]$$

2) In the mercury, one descends from h_3 in order to increase the pressure from the atmospheric pressure p_a to the pressure p_{HG0} to be determined:

$$p_{hgo} = p_a + \rho_{hg} g h_3$$

$$p_0 = 1{,}231.10^5 \; Pa \qquad [2.32]$$

3) In the oil, one descends from h_2 in order to increase the pressure from p_{HG0} to the pressure p_{0W} to be determined:

$$p_{ow} = p_{hgo} + \rho_o g h_2$$

$$p_{hgo} = p_a + \rho_{hg} g h_3$$

$$p_{ow} = p_a + \rho_o g h_2 + \rho_{hg} g h_3 \qquad [2.33]$$

4) In the water, one "descends from X" in order to increase the pressure from p_{OW} to the pressure at the bottom, which is still that which was calculated in eq [2.31], p_0:

$$p_0 = p_{ow} + \rho_w g X \qquad [2.34]$$

$$p_{ow} = p_a + \rho_o g h_2 + \rho_{hg} g h_3$$

$$p_0 = p_a + \rho_w g h_1 + \rho_o g h_2 \qquad [2.35]$$

$$p_a + \rho_w g h_1 + \rho_o g h_2 = p_a + \rho_o g h_2 + \rho_{hg} g h_3 + \rho_w g X$$

$$X = \frac{\rho_w h_1 - \rho_{hg} h_3}{\rho_w} \qquad [2.36]$$

5) For h_3 which is equal to 4 cm, X emerges at:

$$X = 0,456\,m = 45,5\,cm \qquad [2.37]$$

2.2.2. *Case of volume forces deriving from a potential*

EXAMPLE 2.6 (Accelerometer).–

1) In Figure 2.9, a parallelepiped-shaped aquarium is accelerated from left to right in a horizontal direction. Find the shape of the free surface, as well as the geometry of the isobars inside the aquarium.

2) The diagram shows an accelerometer having a U-shaped tube filled with a liquid density ρ. The distance between the vertical arms of the U-shaped tube is $l = 30\,cm$.

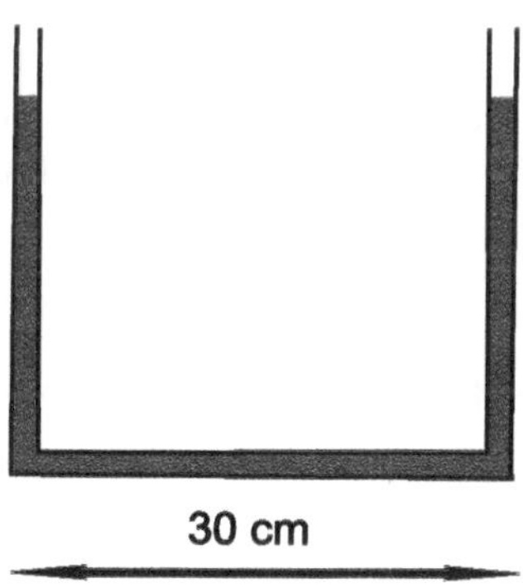

Figure 2.9. *U-shaped tube of the accelerometer*

Explain how one can use this device to measure the horizontal acceleration of the U-tube. Give the relation between the drop observed and the acceleration γ to be measured.

3) One installs the accelerometer into a car.

What gap will a driver observe between the two arms when driving along a district road at 109 km.h^{-1}, given that their GPS foresees that the driver has reached a "danger zone" and wants to bring the speed back down to the legal limit of 90 km.h^{-1} in 5 s? The deceleration is assumed to be constant.

Solution:

1) The problem could be dealt with in statics if one were to place it in a reference frame linked to the aquarium. This frame will be in acceleration in relation to the ground, so it will not be Newtonian.

The volume forces will therefore include a gravitational upward component and an inertia force that has a horizontal direction.

One will therefore choose an axis $Oxyz$ system, integrating a horizontal axis (e.g. Ox), a vertical upward axis, (e.g. Oz) and a third axis with a minor role (e.g. Oy), which is perpendicular to the figure. (The vertical upward axis is constant throughout all of our approaches.)

Both forces (gravity and inertia) are derived from a potential. The resultant force will be derived from the sum of these two potentials:

$$\vec{F}_G = -grad\ \phi_G \qquad [2.38]$$

$$\vec{F}_{inert} = -grad\ \phi_{inert} \qquad [2.39]$$

Taking into account the fact that the inertia force goes in the opposite direction to movement and is carried by Ox, [2.38] and [2.39] allow us to write:

$$\begin{aligned} 0 &= -\frac{\partial \phi_G}{\partial x} \\ 0 &= -\frac{\partial \phi_G}{\partial y} \\ -g &= -\frac{\partial \phi_G}{\partial z} \end{aligned} \qquad [2.40]$$

$$-\gamma = -\frac{\partial \phi_{inert}}{\partial x}$$
$$0 = -\frac{\partial \phi_{inert}}{\partial y} \quad [2.41]$$
$$0 = -\frac{\partial \phi_{inert}}{\partial z}$$

Afterwards, the elementary integrations emerge:

$$\phi_G = gz + C_1 \quad [2.42]$$

$$\phi_{inert} = \gamma x + C_2 \quad [2.43]$$

C_1 and C_2 are constants.

Finally, the potential of the volume forces ϕ, defined at the constant C, is around:

$$\phi = \phi_G + \phi_{inert} = \gamma x + gz + C \quad [2.44]$$

The equipotentials will have the equation:

$$\gamma x + gz = Cte \quad [2.45]$$

These are parallel levels whose track is in the figure's level and have the right equation:

$$z = -\frac{\gamma}{g}x + Cte' \quad [2.46]$$

Cte and Cte' are constants.

We see that the fluid tends to move back (be sent back) toward the rear of the aquarium compared to the movement. This is the classic phenomenon of the passenger being pressed back into their seat in an accelerating vehicle.

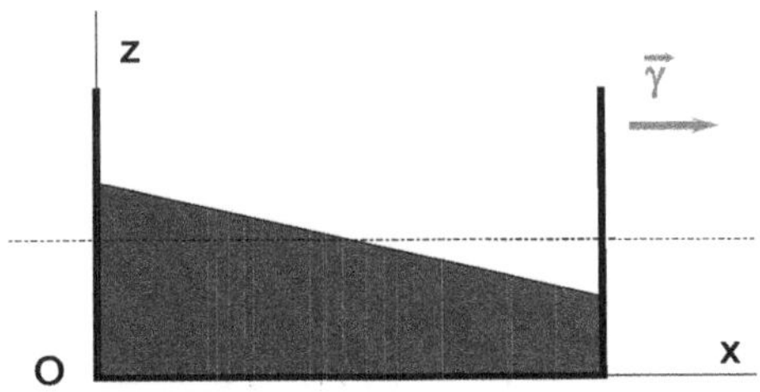

Figure 2.10. *The aquarium in acceleration*

The exact position of the interface may be found by conserving the water volume in the vessel.

2) The following questions are comprised of simple applications of these results.

The shape of the aquarium in no way modifies the previous reasoning. The free surface maintains the same equation and is reduced into two levels within the tube's two vertical sections. Δz is the drop observed. It corresponds to two distant z point values on the interface of $\Delta x = l$.

Therefore:

$$\Delta z = -\frac{\gamma}{g}\Delta x = -\frac{\gamma l}{g} \qquad [2.47]$$

3) The vehicle's deceleration is 19 km.h^{-1} (or 5.28 ms^{-1}) in 5 s, or:

$$\gamma = \frac{19000}{3600 * 5} = 1,056\, m.s^{-2} \qquad [2.48]$$

Δz emerges at $\Delta z = 21,41\, cm$ [2.49]

EXAMPLE 2.7 (Rotating vessel).–

Let us consider a cylindrical receptacle with diameter $D = 2R$ and height H, containing a liquid with a density ρ. This liquid's free surface is at a distance of h_0 $h_0 < H$ from the bottom when the vessel is in recovery.

One puts this vase so that it is rotating around its vertical axis. The angular speed of rotation is ω, expressed in Rd.s^{-1}.

Let us look for the form that the free surface takes on with this rotation.

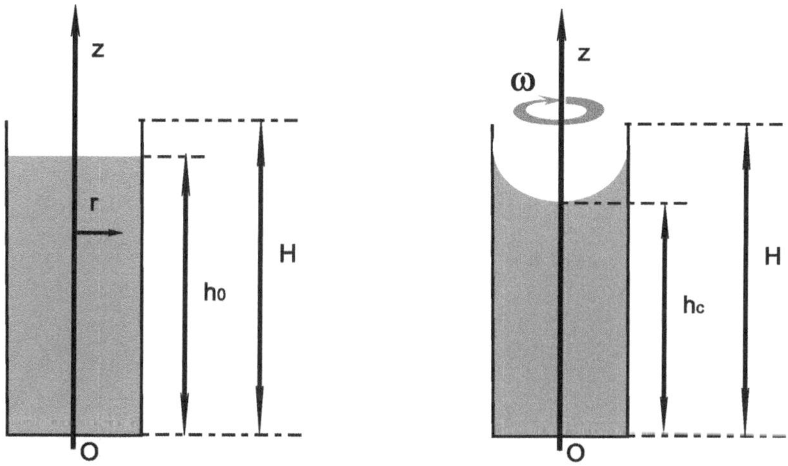

Figure 2.11. *Immobile vessel and rotating vessel*

Consider a cylindrical axis system, r, θ, z , attached to the vessel. In such a frame, the problem arises from fluid statics.

1) Find the free surface's equation, which will naturally be expressed under the form $z = f(r)$. We will call h_C the dimension of the lowest point on this free surface.

2) Express h_C according to h_0 and ω. At which value of ω will the liquid overflow?

3) We have $D = 7\,cm,\ \ H = 10\,cm,\ \ h_0 = 6\,cm$. At which rotating speed does the liquid overflow? Express this speed in mn^{-1} turns.

4) Give the general expression of the pressure in all of the liquid $p = p(r, z, \omega)$.

Solution:

1) The free surface is an isobar, so an equipotential, whose general form will be given by the potential ϕ of the volume forces.

The approach will be the same as in Example 2.6. Nevertheless, one will work in cylindrical coordinates.

The inertia forces will be radial and in a point (r, θ, z) of intensity $r\omega^2$ per mass unit.

If ϕ_G and ϕ_{inert} are the points relating to the forces of gravity and inertia:

$$\phi = \phi_G + \phi_{inert} \tag{2.50}$$

$$\begin{aligned} 0 &= -\frac{\partial \phi_G}{\partial r} \\ 0 &= -\frac{1}{r}\frac{\partial \phi_G}{\partial \theta} \\ -g &= -\frac{\partial \phi_G}{\partial z} \end{aligned} \tag{2.51}$$

$$\begin{aligned} r\,\omega^2 &= -\frac{\partial \phi_{inert}}{\partial r} \\ 0 &= -\frac{1}{r}\frac{\partial \phi_{inert}}{\partial \theta} \\ 0 &= -\frac{\partial \phi_{inert}}{\partial z} \end{aligned} \tag{2.52}$$

Equations [2.51] and [2.52], which are in cylindrical coordinates, correspond to equations [2.40] and [2.41], which are written in Cartesian coordinates.

Then, elementary integrations emerge from this:

$$\phi_G = gz + C_1 \tag{2.53}$$

$$\phi_{inert} = \frac{-r^2\,\omega^2}{2} + C_2 \tag{2.54}$$

C_1 and C_2 are constants.

Finally, the potential of volume forces ϕ, defined at the constant C, is near:

$$\phi = \phi_G + \phi_{inert} = gz - \frac{r^2\,\omega^2}{2} + C \tag{2.55}$$

The equipotentials will have the equation:

$$z = \frac{r^2\,\omega^2}{2\,g} + Cte \tag{2.56}$$

The equipotentials are paraboloids with parallel rotations, whose track in the figure's level is a parabola with the equation:

$$z = \frac{r^2 \omega^2}{2g} + Cte \tag{2.57}$$

2) The equation of the free surface will be found by conserving the fluid's mass.

When the cylinder does not turn, the free surface is a level located at h_0 above the bottom. The liquid's volume is therefore:

$$V_{ol} = \pi R^2 h_0 . \tag{2.58}$$

When the cylinder turns, the fluid is contained in the cylinder that is limited at the bottom, the cylinder's walls and a revolving paraboloids equation revolution:

$$z = \frac{r^2 \omega^2}{2g} + h_C \tag{2.59}$$

where h_C is the height of the liquid at the center ($r = 0$).

The height of the liquid at the level of the wall ($r = R$) will therefore be h_p such as:

$$h_p = \frac{R^2 \omega^2}{2g} + h_C \tag{2.60}$$

In order to calculate the volume of the rotating liquid, we use a classic method for calculating an integral volume with rotating symmetry. We divide the volume of the liquid into small annular spaces held between the cylinder radii r and $r + dr$. Each elementary volume constructed in this way will have the value:

$$dV_{OL} = z(r) 2\pi r dr = \left[\frac{r^2 \omega^2}{2g} + h_C \right] 2\pi r dr \tag{2.61}$$

By integrating 0 to R, the volume can therefore be written as:

$$V_{OL} = \int_0^R \left[\frac{r^2 \omega^2}{2g} + h_C \right] 2\pi r \, dr \tag{2.62}$$

$$V_{OL} = 2\pi \left(\frac{r^4 \omega^2}{8g} + h_C \frac{r^2}{2} \right) \Bigg|_0^R = \frac{\pi R^4 \omega^2}{4g} + \pi h_C R^2 \tag{2.63}$$

Finally, by comparing the two volume expressions, we have:

$$V_{OL} = \frac{\pi R^4 \omega^2}{4g} + \pi h_C R^2 = \pi h_0 R^2 \qquad [2.64]$$

For h_C , it then emerges as:

$$h_C = h_0 - \frac{R^2 \omega^2}{4g}$$
$$h_P = h_0 + \frac{R^2 \omega^2}{4g} \qquad [2.65]$$

The liquid will overflow once h_p is larger than H, which for ω is:

$$h_p = \frac{R^2 \omega^2}{4g} + h_0 > H \qquad [2.66]$$

$$\omega^2 > \frac{4g(H - h_0)}{R^2} \qquad [2.67]$$

3) We have $D = 7\,cm$, $H = 10\,cm$, $h_0 = 6\,cm$.

The digital application is given by:

$$\omega = 35{,}8\,Rd.s^{-1}, \qquad [2.68]$$

$$\omega = \frac{35{,}8}{2\pi} * 60 = 341{,}9.mn^{-1} turns \qquad [2.69]$$

for the speed given above at which the liquid overflows.

4) Let us look for the general expression of the pressure in all of the liquid $p = p(r,z,\omega)$.

The isobars have the general equation:

$$z = \frac{r^2 \omega^2}{2g} + h_C = \left[h_0 - \frac{R^2 \omega^2}{4g}\right] + \frac{r^2 \omega^2}{2g} \qquad [2.70]$$

We have an incompressible fluid with a force potential:

$$\phi = \phi_G + \phi_{inert} = gz - \frac{r^2 \omega^2}{2} + C \qquad [2.71]$$

So, according to [2.6]:

$$p + \rho\phi = p + \rho\left(g\,z - \frac{r^2 \omega^2}{2}\right) = K \qquad [2.72]$$

where K is a constant that one will assess when writing that the free surface is an isobar with the value p_a. This is particularly the case for $z = h_C$ and $r = 0$. Therefore:

$$\begin{aligned} &p_a + \rho(gh_C) = K \lim_{x \to \infty} \\ &K = p_a + \rho(gh_C) = \rho g\left(h_0 - \frac{R^2 \omega^2}{4g}\right) \end{aligned} \qquad [2.73]$$

Therefore, the general expression for the pressure in the liquid is:

$$p = K - \rho\left(gz - \frac{r^2 \omega^2}{2}\right) = \rho g\left(h_0 - \frac{R^2 \omega^2}{4g}\right) - \rho\left(gz - \frac{r^2 \omega^2}{2}\right) \qquad [2.74]$$

$$p\,(r, z, \omega) = p_a + \rho g\,(h_0 - z) + \rho\left(\frac{r^2 \omega^2}{2} - \frac{R^2 \omega^2}{4}\right) \qquad [2.75]$$

2.2.3. *Case for compressible fluids*

Once the fluid can no longer be considered incompressible, which is particularly the case for gases, the fluid statics fundamental theorem cannot be applied. We therefore need to use the formulations given in 2.1.1.b

NOTE.– Other liquids may also be compressible if one considers significant pressure variations. For example, at a depth of 10,000 m, the pressure varies by about 100 bars; the liquid's compressibility should therefore be considered less.

EXAMPLE 2.8 (The evolution of atmospheric pressures with altitude).–

Air pressure in the atmosphere varies considerably depending on the altitude.

We will attempt to model how the pressure evolves with altitude by assuming that the air is a perfect gas and by using two hypotheses:

a) The atmosphere is isothermal. Any mountain climber or aviator will be able to confirm that this is a fairly rough hypothesis.

b) One atmospheric layer passes to another layer within the context of reversible adiabatic transformation. This hypothesis is more realistic.

So p_0 and ρ_0 are the pressure and the density from the air to the ground, respectively.

We recall that for a perfect gas:

The status equation: $\frac{p}{\rho} = rT$ is verified with $r = \frac{R}{M}$, the ratio of the constant for perfect gases to molar mass.

NOTE.– r appears as soon as one starts to work in mass flow, which is often the case in compressible fluid mechanics. Note that unlike R, r is not a universal constant, but instead varies with the nature of the gas.

T is the absolute temperature of the gas; therefore, it is expressed in Kelvin.

In a reversible adiabatic transformation (isentropic), the quantity $\frac{p}{\rho^\gamma}$ remains constant. $\gamma = \frac{C_P}{C_V}$, the ratio of the heat capacities to the pressure and constant volume, is also called a polytrophic coefficient of the gas.

1) Give the expression of the atmospheric pressure $p(z)$ in the case of an isothermal atmosphere, where z is the altitude.

2) Give the expression $p(z)$ in the hypothesis of a reversible adiabatic transformation.

3) We have $p_0 = 1,013.10^5\ Pa$; $\rho_0 = 1,29\,kg.m^{-3}$; $\gamma = 1,4$.

What is the pressure at the altitude of 10,000 m for each hypothesis?

Solution:

We are now dealing with compressible fluids. The fluid statics law is no longer suitable.

We bring the space back to the frame that possesses a vertical upward axis Oz, where the origin is fixed at the ground surface.

We have to refer back to the general expression of the fluid statics theorem, in the case of a volume force reduced to gravity, therefore derived from a potential $\phi = -g\,z$ per mass unit:

$$\vec{grad}\,p + \rho\,\vec{grad}\,\phi = 0 \qquad [2.76]$$

Let us project this principle onto the vertical axis Oz:

$$\frac{\partial p}{\partial z} = -\rho\frac{\partial \phi}{\partial z} \qquad [2.77]$$

We can rewrite this expression here using "total" derivatives as the relation of gradients enables us to demonstrate that the isobar network ($p = Cte$) is confounded with that of the equipotentials (in this case, they are levels at $z = Cte$).

The pressure p, like ϕ, only varies with z:

$$\frac{dp}{dz} = -\rho\frac{d\phi}{dz}$$

$$\frac{dp}{dz} = -\rho g \qquad [2.78]$$

We can write a supplementary ratio between p and ρ. This ratio will depend on the thermodynamic transformation of the air moved from one altitude to another.

1) If we assume that the temperature of the atmosphere is constant, the following state equation will be verified at each dimension:

$$\frac{p}{\rho} = rT = \frac{p_0}{\rho_0} \qquad [2.79]$$

T is the absolute temperature of the gas. In this first case, rT is therefore a constant. Small variations in pressure and density will be connected by:

$$dp = \frac{p_0}{\rho_0} d\rho \qquad [2.80]$$

In this particular case, the fundamental theorem of statics will be written as:

$$\frac{dp}{dz} = -g \frac{\rho_0}{p_0} p \qquad [2.81]$$

$$\frac{dp}{p} = -\, g \frac{\rho_0}{p_0} dz \qquad [2.82]$$

This differential equation of the first range is easily integrated. The condition on the necessary limits will be given by the value of atmospheric pressure at the ground p_a:

$$Ln\ p = -\, g \frac{\rho_0}{p_0} z + Ln\, C \qquad [2.83]$$

$$p = C \exp -\, g \frac{\rho_0}{p_0} z$$

$$z = 0 \ ; \ p = p_0$$

$$p = p_0 \exp -\, g \frac{\rho_0}{p_0} z \qquad [2.84]$$

This is an exponential decrease in the pressure with the temperature.

This is obviously a very simplified model. Indeed, we know that temperature at altitudes is significantly lower than the temperature at the ground level.

2) We will therefore obtain a better approximation by assuming a reversible adiabatic relation between the two points of the different dimensions:

$$\frac{p}{\rho^{\gamma}} = K$$

$$\gamma = \frac{c_P}{c_V} \qquad [2.85]$$

where K is a constant.

By introducing this new ratio [2.85] and density in the fundamental theorem of statics, we will obtain a new differential equation of the first range for p:

$$\frac{p}{\rho^{\gamma}} = K = \frac{p_0}{\rho_0^{\gamma}}$$

$$\rho = \left(\frac{p}{K}\right)^{\frac{1}{\gamma}} = \left(\frac{\rho_0^{\gamma}}{p_0}\right)^{\frac{1}{\gamma}} p^{\frac{1}{\gamma}} = \frac{\rho_0}{p_0^{\frac{1}{\gamma}}} p^{\frac{1}{\gamma}} \qquad [2.86]$$

$$\frac{dp}{dz} = -\rho g$$

It appears as:

$$\frac{dp}{p^{\frac{1}{\gamma}}} = -\frac{\rho_0}{p_0^{\frac{1}{\gamma}}} g\, dz = -A\, dz$$

$$A = \frac{\rho_0}{p_0^{\frac{1}{\gamma}}} g \qquad [2.87]$$

Equation [2.87] can be solved by using the same condition on the limits as before: $z = 0$; $p = p_a$

$$\frac{\gamma}{\gamma-1} p^{\frac{\gamma-1}{\gamma}} = -Az + Cte$$

$$\frac{\gamma}{\gamma-1} p^{\frac{\gamma-1}{\gamma}} = -Az + \frac{\gamma}{\gamma-1} p_0^{\frac{\gamma-1}{\gamma}} = -Az + B$$

p emerges at:

$$p = \left[\frac{\gamma-1}{\gamma}(-Az + B)\right]^{\frac{\gamma}{\gamma-1}} \qquad [2.88]$$

where

$$A = \frac{\rho_0}{p_0^{\frac{1}{\gamma}}} g$$

$$B - \frac{\gamma}{\gamma-1} p_0^{\frac{\gamma-1}{\gamma}}$$

3) The values are $p_0 = 1,013.10^5\ Pa$; $\rho_0 = 1,29\,kg.m^{-3}$; $\gamma = 1,4$.

3.1) In the hypothesis of an isothermal atmosphere:

$$p = p_0 \exp - g\frac{\rho_0}{p_0} z$$

$$p = 1,013.10^5 \exp - 9,81\frac{1,29}{1,013.10^5} 10^4$$

$$p = 2,9.10^4\ Pa = 0,29\,bar \qquad [2.89]$$

3.2) In the hypothesis of a reversible adiabatic evolution with altitude:

$$p = \left[\frac{\gamma-1}{\gamma}\left(-A\ z + B\right)\right]^{\frac{\gamma}{\gamma-1}}$$

$$A = \frac{\rho_0}{p_0^{\frac{1}{\gamma}}} g = 3,364.10^{-3}$$

$$B = \frac{\gamma}{\gamma-1} p_0^{\frac{\gamma-1}{\gamma}} = 94,24$$

$$p = \left[0,286\ (-33,64 + 94,24)\right]^{3,5}$$

$$p = 2,17.10^4\ Pa = 0,217\,bar \qquad [2.90]$$

The International Organization for Standardization (ISO) published the International Standard Atmosphere (*ISA*) under the standard, ISO 2533; 1975. We can compare the result of this data to the standard ISA atmosphere, which, in the troposphere at the sea level, gives a pressure of $1,013.10^5\ Pa$ at an altitude of $11000\ m$ and gives a standardized pressure of $22\,632\,Pa$. Assuming that the pressure has an approximately linear evolution in the troposphere, we can find:

$$p = 0,226\frac{10}{11} = 0,206\,bar \qquad [2.91]$$

This result can be compared with the previous ones.

The isothermal approximation gives a gap of 40.1%, while the isentropic approximation gives a reduced gap of 5.3% with the standard data.

2.3. Calculating the thrusts

2.3.1. *Methods*

The thrust that a fluid exerts on a finite surface S, whether open or closed, is in principle calculated by a surface integral:

$$\vec{P}_{thrust} = \iint_S -p(z)\vec{n}\, dS \tag{2.92}$$

where $p(z)$ will be a function obtained based on the fluid statics fundamental theorem, under a more or less complex form.

There are diverse symmetries (the simplest of which are the mono-dimensional plane and the axis symmetry) that often simplify the calculations.

In the most general case, it may lead to particularly complex calculations. In practice, complex calculations call upon the digital approach.

For a closed surface (e.g. an immobile immersed solid body), in fluid statics, we can benefit from using Archimedes' theorem.

Hereafter, the equilibrium or the movement of the solid body is a problem in solid mechanics. The thrust calculated on the closed surface enveloping this body can therefore only be a component in a system of forces.

NOTE.– In the case of a moving body, applying the Archimedes theorem bypasses the immobility of the solid. This remains acceptable for slow-speed movements: sedimentation, ascension of a hot air balloon, etc.

This encompasses different issues: buoyancy, sedimentation, ascension of an airship, etc.

Two types of methods will be broached below:

a) Thrust calculations on closed and immersed surfaces.

b) Thrust calculations on an open surface (wall).

Calculating the thrust on an open surface leads to the decomposition of the surface into elementary components of relevant geometry. In practice, we look for the symmetries and often use a band or ring decomposition.

2.3.2. *Thrusts on bodies that are totally immersed in incompressible fluids*

For incompressible fluids subject to the only gravity, Archimedes' theorem is a practical tool for calculating thrusts on a closed surface.

All bodies immersed in an incompressible fluid that is subject to volume forces reduced to gravity receive a vertical upward thrust from this liquid, with an intensity equal to the shifted fluid weight. This thrust is applied to the center of the mass of the displaced fluids.

It is very important to understand that Archimedes' theorem is just one way of calculating a thrust applied by a fluid responding to hypotheses. The problems that appeal to this theorem are furthered by studying the equilibrium of the body itself. From here, we leave fluid mechanics and enter into the realm of solid mechanics; this is where we find buoyancy problems, in particular.

NOTE.– Let us recall that incompressibility is synonymous with *invariable density*. Incompressible fluids therefore include liquids subject to reasonable pressure (at the bottom of the Pacific, the liquid is compressed) and gases subject to weak pressure variations.

There are several observations to take into account here:

Archimedes' theorem implies immobile bodies immersed into an incompressible fluid subject to volume forces reduced to gravity.

This theorem does not apply to liquids. It can be applied to a gas, on the condition that its density does not sensitively vary in the volume in which one is dealing with the problem.

In the case where the immersed body is slowly moving, we assume that, for the purposes of estimation, Archimedes' theorem can still evaluate the thrust forces of the fluid. This estimate is used particularly for the movement of an aerostat (see the hot air balloon in Example 2.13) or in sedimentation problems.

NOTE.– Sedimentation time from a few minutes to a few hours can be observed when a particle goes down to a few centimeters.

EXAMPLE 2.9. (A floating ball).–

A ball with a diameter D is full and made up of a homogeneous material with a density ρ_B.

We put this ball on the surface of a tank filled with water. The water density is ρ. We know that $\rho_B < \rho$.

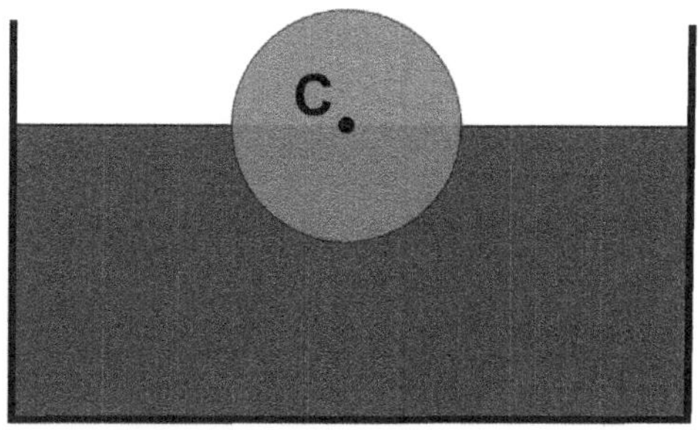

Figure 2.12. *A floating ball*

1) In order for the ball to be immersed in such a way that its center C is at the level of the tank's free surface, what should the value of ρ_n be ρ_B? In this question, we will only give the expression.

2) We have $\rho = 1000\ g.cm^{-3}$.

Give the value of ρ_B in SI units.

Solution:

This simple example enables us to find the basis of the logic that will be reproduced in practically all the problems which call upon Archimedes' thrust.

At this stage, do not forget that we are indeed dealing with a solid's statics, where fluid mechanics only occurs for calculating a force's components, the thrust.

1) The fact that the sphere's center is at the free surface level implies that half of the sphere is only immersed.

V_{OL} is the sphere's volume: $V_{OL} = \dfrac{\pi d^3}{6}$, where d is the diameter (which is unknown, but we will see that this is not necessary to know).

The sphere is immobile. Therefore, it is in equilibrium under the action of two forces, an upright support in the opposite direction:

Gravity force, with intensity F_G is the sphere's weight.

The liquid's thrust, which by averaging the fulfilled hypotheses, can be calculated as an Archimedes' thrust, with intensity F_A is equal to the shifted liquid weight. In this problem, the shifted liquid volume is half of the sphere's volume

$$V_{imm} = \frac{\pi d^3}{12} \quad [2.93]$$

The equilibrium, projected on a vertical upward axis, is written as:

$$\begin{aligned} &F_A - F_G = 0 \\ &\rho V'_{OL}\, g = \rho_B\, V_{OL}\, g \\ &\rho \frac{\pi d^3}{12} g = \rho_B \frac{\pi d^3}{6} g \end{aligned} \quad [2.94]$$

As a result, we can obtain a simple relation:

$$\rho_B = \frac{\rho}{2} \quad [2.95]$$

We see that the diameter of the sphere is eliminated in the calculation.

2) The digital value of ρ_B immediately becomes:

$$\rho_B = \frac{\rho}{2} = 500\, g.cm^{-1} = 500\, kg.m^{-3} \quad [2.96]$$

EXAMPLE 2.10 (A bear on the ice floe).–

A 16 kg polar bear is sitting on a piece of ice whose dimensions are unknown value. The ice is barely immersed and immobile.

The respective densities of the seawater and the ice are $\rho = 1025\, kg.m^{-3}$ and $\rho_S = 922\, kg.m^{-3}$.

What is the volume V_{OL} of the piece of ice floe?

Solution:

The piece of ice floe must be at the equilibrium under the action of forces applied to it, namely:

the weight of the ice added to the weight of the bear are directed downwards;

Archimedes' theorem, directed upwards, whose intensity is equal to the weight of the shifted water volume. This volume here is equal to the ice volume that is totally immersed.

These forces are in a vertical direction (carried by a vertical axis) and in the opposite direction. The equilibrium is written as:

$$(m + \rho_S\, V)\, g = \rho V g \qquad [2.97]$$

As a result, we obtain the volume V_{OL} of the piece of ice floe:

$$V_{OL} = \frac{m}{\rho - \rho_S} \qquad [2.98]$$

$$V_{OL} = 1,55\, m^3 \qquad [2.99]$$

EXAMPLE 2.11 (A tourist in the Dead Sea).–

The Dead Sea contains 44 g of salt per liter of water. An English tourist with a density ρ_T of 0.9 g.cm^{-3} floats. In volume percentage, what fraction f of his body is immersed? We consider that completely desalinated water has a density of $\rho = 1000\, kg.m^{-3}$.

Solution:

V_{OL} is the volume, which is still unknown to the tourist.

The tourist must be at equilibrium under the action of the forces that are applied on him, namely:

his weight $\rho_T\, V_{OL}\, g$;

Archimedes' upward thrust, whose intensity is equal to the weight of the displaced water volume. This volume here is equal to the immersed part of the tourist's body $f\, V_{OL}$.

These forces are in a vertical direction (carried by a vertical axis) and in the opposite direction. The equilibrium is written as:

$$\rho_T V_{OL} g = \rho f V_{OL} g \qquad [2.100]$$

The volume is removed and:

$$f = \frac{\rho_T}{\rho} \qquad [2.101]$$

The salt water density is the sum of the water and salt mass in m^3 ($44\, g.litres^{-1} = 44\, kg.m^{-1}$).

The digital value of f is:

$$\rho = 1000 + 44 = 1044\, kg.m^{-3}$$

$$f = \frac{922}{1044} = 0{,}883 = 88{,}3\% \qquad [2.102]$$

EXAMPLE 2.12 (Thrust on an immersed pebble).–

A physicist wants to know the thrust that the sea applies onto a pebble on the Dieppe beach. As the pebble has a very irregular shape, he abandons a direct calculation.

He fills a cylindrical vessel of diameter $D = 10\, cm$ with water. He plunges this pebble into the vessel, where it then becomes completely immersed. The water level rises by $\Delta z = 9\, cm$.

When the pebble is totally immersed, what is the thrust applied by the sea onto this pebble?

The density of the seawater is $\rho = 1025\, kg.m^{-3}$.

Solution:

Archimedes' theorem outlines that the thrust undergone by a body immersed in a liquid is vertically upward and has an intensity equal to the weight of the shifted fluid. It is also applied to the center of the mass of this shifted fluid.

The rise in the water level enables us to calculate the volume of the pebble. This volume is given by:

$$V_{OL} = S\ \Delta z = \pi \frac{D^2}{4} \Delta z$$
$$V_{OL} = 7{,}85.10^{-3} * 9.10^{-2} = 7{,}07.10^{-4}\ m^3 \qquad [2.103]$$

The intensity of the thrust F, equal to the weight of the shifted fluid, is therefore:

$$F = \rho\, V_{OL}\ g = S\ \Delta z = \pi \frac{D^2}{4} \Delta z \qquad [2.104]$$

$$F = 1025 * 7{,}07 * 10^{-4} * 9{,}81 - 7{,}108\ N \qquad [2.105]$$

EXAMPLE 2.13. (A historic event).–

On June 23rd, 1784, Pilâtre de Rozier flew aboard a hot air balloon, named the "Marie-Antoinette", which was built by the Montgolfier brothers. Throughout this problem, we will assume that a hot air balloon is formed by a perfectly spherical envelope, with a diameter d, containing hot air with a density ρ_C.

ρ_C is less than the density ρ of the air surrounding the hot air balloon.

Let m be the total mass of the envelope, the basket and the pilot. m does not include the mass of the air contained in the envelope.

The aerodynamic effects on the basket are insignificant.

The density ρ of the external air decreases with the altitude, in accordance with the law:

$$\rho = \rho_S (1 - \alpha h)$$

where α is a constant coefficient and ρ_S is the density at the ground. ρ_C remains constant

1) Give the literal expression of the altitude H that this flight, one of the first human flights, could have reached?

2) Numerical application

Calculate H knowing that:

$m = 100\,kg$	$\rho_C = 0{,}95\,kg.m^{-3}$	$g = 9{,}81\,m.s^{-2}$
$d = 11{,}5\,m$	$\rho_S = 1{,}3\,kg.m^{-3}$	$\alpha = 8{,}8.10^{-5}$

Solution:

1) At the height H, the hot air balloon is immobile. Therefore, we here have a solid statics problem. The hot air balloon is in equilibrium under the action of the forces that are all carried along a vertical axis (which one will direct upward. The origin will again be at the ground level):

Forces of gravity, with mass m, applied onto the envelope, the basket and the pilot. With an intensity of $F_{Genv} = mg$, these forces will be directed downward.

Forces of gravity applied onto the gas contained in the envelope. With an intensity of $F_{Ggas} = m'g$, where the mass of the gas is $m' = \pi \dfrac{D^3}{6} \rho_C$, these forces are directed downward.

Thrust of ambient air, external to the envelope. This force has an intensity equal to the weight of the shifted air, or an intensity $P = m_{air}\, g$. It is directed upward.

The shifted air has a mass:

$$m_{air} = \pi \frac{D^3}{6} \rho = \pi \frac{D^3}{6} \rho_S \left(1 - \alpha h\right) \quad [2.106]$$

This last force only varies with the altitude h.

At equilibrium, $h = H$. This equilibrium is written, projected on a vertical upward axis:

$$m_{air}\, g - \left(m + m'\right) g = 0 \quad [2.107]$$

$$m_{air} = m + m'$$

$$\pi \frac{D^3}{6} \rho_S \left(1 - \alpha H\right) = m + \pi \frac{D^3}{6} \rho_C \quad [2.108]$$

In this equality, H is the parameter to be determined. Let us isolate:

$$H = \frac{1}{\alpha\rho_S}(\rho_S - \rho_C) - \frac{6m}{\pi D^3} \quad [2.109]$$

2) Numerically, H is calculated as:

$$H = 1962\,m \quad [2.110]$$

EXAMPLE 2.14 (Continental isostasy).–

The continents are made up of a crust with a density ρ_c and a constant thickness in the time h. This crust "floats" on the surface of a relatively fluid environment, which we will very boldly assimilate here to a *liquid* with a density ρ_m.

We will look at the case of Scandinavia.

At the beginning of the quaternary era, Scandinavia was covered with a thick layer of ice, ice sheets or polar cap, with a thickness e. The density of this ice sheet is ρ_G. The surfaces of the ice and the crust are identical.

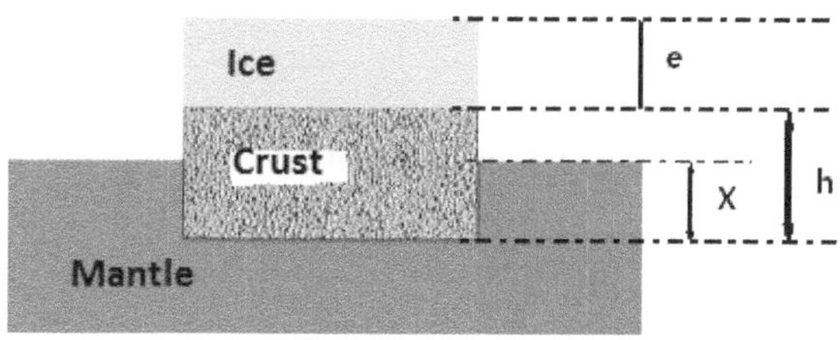

Figure 2.13. *Simplified diagram of Scandinavia*

1) At what depth X was the crust immersed at the beginning of our era?

2) One knows that in T years, the ice sheet had melted; this melting causes Scandinavia to regularly rise at a constant speed U_R. Now, there is practically nothing more than a negligible thickness of ice. What is the depth X' at which Scandinavia is currently?

3) *Numerical application*: expressed in meters per decade, at which speed U_R has Scandinavia risen?

$$\rho_C = 2900\,kg.m^{-3};\ \rho_M = 3300\,kg.m^{-3};\ \rho_G = 990\,kg.m^{-3}$$

$$e = 2000\,m;\ h = 36\,km;\ T = 60\,000\,ans$$

Solution:

1) At the beginning of the quaternary era, Scandinavia was at equilibrium under the action of the forces that were all carried along a vertical axis (which we will direct upward, originating at the ground level). We will record Scandinavia's air as S:

– forces of gravity, applied onto the crust, with a mass $m_C = \rho_C\, S\, h$ and the ice, with a mass $m_G = \rho_G\, S\, e$. These forces are directed downward.

– mantle's thrust: this force, which is directed upward, has an intensity equal to the weight of the shifted mantle, which has a mass $m_M = \rho_M\, SX$.

The equilibrium is written as:

$$m_M\, g - (m_C + m_G) g = 0 \qquad [2.111]$$

$$\rho_M\, SXg = (\rho_C\, Sh + \rho_G\, Se) g \qquad [2.112]$$

X is calculated as:

$$X = \frac{\rho_C\, h + \rho_G\, e}{\rho_M} \qquad [2.113]$$

2) The ice has disappeared, which amounts to redoing the previous question with $e = 0$.

X' therefore is calculated as:

$$X' = h\frac{\rho_C}{\rho_M} \qquad [2.114]$$

3) Knowing X and X', U_R is given by:

$$U_R = \frac{X - X'}{T} \qquad [2.115]$$

Using the previous results, X and X' can be determined digitally:

$$X = \frac{\rho_C\, h + \rho_G\, e}{\rho_M} = 32\,236\, m$$
$$X' = h\frac{\rho_C}{\rho_M} = 31\,636 \qquad [2.116]$$

U_R is equal to:

$$U_R = \frac{32236 - 31636}{60000} = 10^{-2}\ m.an^{-1} \qquad [2.117]$$

U_R is 1 m per century.

Note that we can find X–X' more directly from their expressions. Knowing h is therefore not necessary.

EXAMPLE 2.15 (Study of a photophore).–

A photophore is a device found in furniture shops, a diagram of which is shown below.

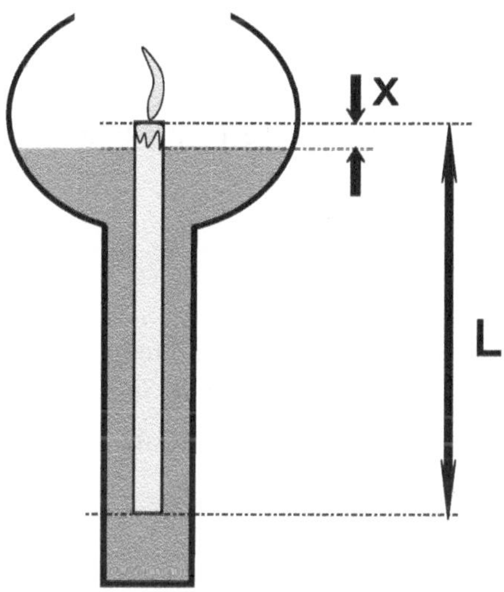

Figure 2.14. *A photophore*

The candle B is partially immersed into the water with a density ρ. The density of the candle is ρ_B. The candle burns at a speed U_f. This speed is defined by the changing length of the candle $L(t)$:

$$U_f = -\frac{dL}{dt}$$

We designate x (t) as the length of the candle emerging from the water.

x_0 is the length of the emerged candle when the candle is lit.

1) Find the literal expression of x(t). One assumes that the candle burns slowly, and therefore all the candle's movements are extremely slow.

2) We have: $\rho_B = 850\, kg.m^{-3}$; $\rho = 1000\, kg.m^{-3}$; $L = 10\, cm$; $U_f = 5\, cm.hr^{-1}$.

By how much does the candle emerge when it is lit? By how much will it have emerged after the candle has burned for 1 hour and 30 minutes? What then, is the total length of the candle?

3) In your opinion, what is the use of such a utensil, apart from its aesthetic addition to a table?

Solution:

1) The candle sinks slowly. We could admit that the problem is quasi-static, which enables us to write:

The candle is "practically" at equilibrium under the action of the forces applied to it, namely the force of gravity and the fluid's thrust, both of which are on a vertical axis.

The liquid's thrust will be equal to Archimedes' thrust, which is strictly valid for an immobile fluid.

S is the section (currently unknown, as it shall remain) of the candle. The volume of the candle is $V_B = SL$ its weight $P_B = \rho_B SLg$. The immersed volume is $V_{imm} = S(L - x)$ the weight of the shifted fluid, and the intensity F_A of Archimedes' force is:

$$F_A = \rho V_{imm}\, g = \rho S(L - x)g \qquad [2.118]$$

Archimedean weight and force are vertical and in opposite directions.

We can write the equilibrium of the candle as:

$$\begin{aligned} &F_A - P_B = 0 \\ &\rho S(L - x)g - \rho_B SL\, g = 0 \end{aligned} \qquad [2.119]$$

As a result, we obtain:

$$x = L\frac{\rho - \rho_B}{\rho} \qquad [2.120]$$

2) The free height of the candle therefore depends on the combustion progress.

2.1) When the candle is new:

$$x = L\frac{\rho - \rho_B}{\rho} = 0,1\frac{150}{1000}$$
$$x = 15.10^{-3}\ m = 1,5\,cm \qquad [2.121]$$

2.2) After 1 hour and 30 minutes, the length of the candle is reduced by $U_F\,t = 5.10^{-2}\ .1,5 = 7,5.10^{-2}\ m$.

Or a remaining length of $L' = 0,1 - 7,5.10^{-2} = 2,5.10^{-2}\ m = 2,5\,cm$.

The ratio established between x and L remains valid, regardless of the value of L. So, after 1 hour and 30 minutes, the height of the immersed candle becomes:

$$x' = L'\frac{\rho - \rho_B}{\rho} = 0,025\frac{150}{1000}$$
$$x' = 3,75.10^{-3}\ m = 3,75\,mm \qquad [2.122]$$

3) We observe that, for two very different candle lengths, x conserves a weak value, which, from a practical point of view, ensures that the flame is always close to the liquid's surface, whose position varies very little. In this particular case, a glass globe surrounds the flame to ensure that the light is dispersed or focused ("lamp shade glass" effect). This is an aesthetic effect that is maintained identically throughout the combustion of the candle.

EXAMPLE 2.16. (Alcohols).–

Readers are reminded in advance that alcohol abuse is harmful to your health.

At a party, a physicist sees a guest holding a glass in which an ice cube is floating. He notices that the ice cube is totally immersed and just touches the liquid's surface, where it stays at equilibrium. He wonders what this guest is drinking.

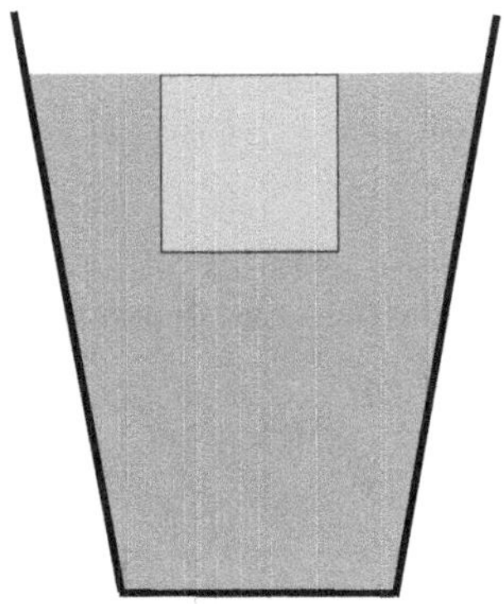

Figure 2.15. *The glass and its ice cube*

1) Give the expression of the density ρ of a water–alcohol solution according to its alcohol content n. ρ_{alc} and ρ_w are, respectively, the densities of ethyl alcohol and water.

2) Determine the liquid's density ρ_{bev} in the guest's glass, and from this deduce the alcohol content and the type of beverage.

3) Alcoholmeter:

A hollow cylinder with a volume V_{OL} has a density ρ_C. (ρ_C is the ratio of the mass of the cylinder envelope and the air contained within divided by its external volume.)

When plunged into Sangria, it is immersed by 40%.

Which fraction *f* of its volume will be immersed in Absinthe?

Data: We will help you out with the following information:

We recall that the alcohol content n (noted as $n°$) of a water/alcohol mix is the proportion *in volume, expressed in a percentage,* of alcohol contained in this mix: for example, a liter of "alcohol" at 45° contains 45% volume alcohol, meaning 450 cm^3 of alcohol and 550 cm^3 of water.

Throughout the problem, we will assume that the commercial beverages exclusively contain ethyl alcohol, and we will ignore the effect that all other ingredients (coloring, tannins, additives, etc.) may naturally (or artificially) have on the density.

Alcoholic content (in °) of a few liquids:

Sangria: 14°;

Vermouth: 16°;

Sake (lower quality): 17°;

Sake (higher quality): 20°;

Whisky: 41°;

Mei-Kwei-Lu: 51°;

Absinthe: 75°.

NOTE.– We often ignore the fact that Mei-Kwei-Lu is a Chinese alcohol that is systematically served at various French restaurants, under the name Saké (a generic Japanese word meaning alcohol).

The densities (ρ, in $kg.m^{-3}$) of the elements present in the problem (pure bodies) are as follows:

Pure water: $\rho_w = 1000\, kg.m^{-3}$;

Ethyl alcohol: $\rho_{alc} = 790\, kg.m^{-3}$;

Ice: $\rho_{ice} = 914\, kg.m^{-3}$.

Solution:

1) The alcohol content is the density of alcohol in a water alcohol solution. Expressed in a percentage, $n°$ therefore means that the alcohol represents $\frac{n}{100}$ of the solution's volume.

The density is determined by adding the evaluated water and alcohol masses contained in a m^3 of solution:

$$\rho = \frac{n}{100}\rho_{alc} + \left(1 - \frac{n}{100}\right)\rho_w \qquad [2.123]$$

2) The ice cube is just at the surface. It is totally immersed, immobile and therefore at equilibrium. Its weight will be equal to Archimedes' thrust. The shifted volume is that of the ice cube, V_{ice} , which we will see that it is not necessary to know:

$$\rho V_{ice}\, g = \rho_{bev} V_{ice}\, g \quad [2.124]$$

$$\rho = \rho_{bev} \quad [2.125]$$

The density of the beverage is equal to that of the ice. Thus, we can deduce $n°$:

$$\rho = \frac{n}{100}\rho_{alc} + \left(1 - \frac{n}{100}\right)\rho_w = \rho_{ice}$$
$$\frac{n}{100}790 + \left(1 - \frac{n}{100}\right)1000 = \rho_{ice} = 914\, kg.m^{-"} \quad [2.126]$$

n can be calculated as:

$$n = 41° \quad [2.127]$$

The beverage is Whisky.

3) We calculate the density of Sangria, ρ_{sang}, with an alcoholic content $n = 14°$ and that of Absinthe, ρ_{abs}, with an alcohol content of $n = 75°$:

$$\rho_{sang} = 0,14\,\rho_{alc} + 0,86\rho_w \; = 995,8\, kg.m^{-3} \quad [2.128]$$

$$\rho_{abs} = 0,75\,\rho_{alc} + 0,25\rho_w \; = 842,5\, kg.m^{-3} \quad [2.129]$$

Under the influence of gravity and Archimedes' thrust, the alcoholmeter is at equilibrium. The volume immersed in the Sangria will be $0,4V_{OL}$, and in the Absinthe, it will be $f\,V_{OL}$:

$$0,4\,\rho_{sang} V_{OL}\, g = \rho_C V_{OL}\, g \quad [2.130]$$

$$f\,\rho_{abs} V_{OL}\, g = \rho_C V_{OL}\, g \quad [2.131]$$

f can easily be calculated as:

$$f = 0,4\frac{\rho_{sang}}{\rho_{abs}}$$
$$f = 47,3\% \quad [2.132]$$

EXAMPLE 2.17 (In the times of the Vikings).–

Calculations will only be performed digitally in Question 4.

A Swedish boat or *Knörr* in the 11th Century had the shape of a prism with the length L whose section is approximated by a triangle with side a and height H.

Knarr or *Knörr* is the correct name for the classic Viking ship. "*Drakkar*" is the name that comes from the Romantic era and has no established linguistic basis.

In order to simply model the phenomenon of buoyancy, let us assume that the sea is ideally calm. Consequently, the axis Oz remains vertical. The mass of the boat when empty is m; it includes the keel, the decks and the masts. ***

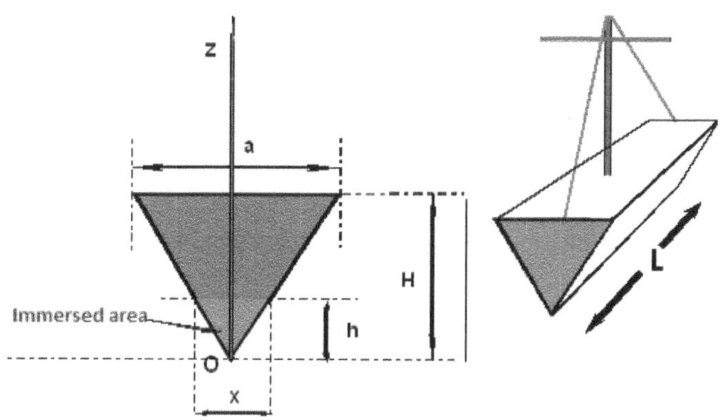

Figure 2.16. *Mapping of a "knörr"*

1) Express the volume of the immersed hull, V_{imm}, according to the wet height h, a, H and L.

2) What is the expression of the immersed hull's height h_0 when the boat has no passengers on board?

3) One now adds a mass M into the boat. Give the relation between h and M. What is the maximum load M_{max} that the boat can transport before sinking?

4) Numerical application

The boat is a *knörr* with length $L = 20\,m$, width $a = 5\,m$ on the upper part of the hull and the hull's height $H = 6\,m$.

When empty, the boat's mass is 10 tons.

4.1) Give the numerical value of h_0.

4.2) Give the numerical value of $M_{\max}$.

4.3) In order to ensure a good pillaging campaign, the following data must be included: 4m^3 of drinkable water, 500kg of provisions and a ton of weapons. Vikings are small, but heavy (they drink lots of beer and mead). Their average mass is evaluated to be 90 kg.

Even in these cold countries, water has a density of 1,000 kg.m^{-3}.

Knowing that we need to provide for rolling and pitching hazards, *only 2/3 of the boat's height can be immersed* when the boat is vertical and resting perfectly.

In theory, how many Vikings can leave to destroy the coasts of Normandy on this atypical *drakkar*? Would it be reasonable to put just as many on such a boat?

Solution:

1) The immersed section and the boat section have similar triangles. We can therefore write:

$$\frac{x}{a} = \frac{h}{H} \qquad [2.133]$$

The immersed volume can immediately be deduced, like the product of the triangle's air immersed by the boat's length:

$$V_{sub} = \frac{hx}{2} L = \frac{a\,h^2\,L}{2\,H} \qquad [2.134]$$

2) The boat is immobile. Therefore, it is at equilibrium under the action of the vertical support forces and in the opposite direction:

the force of gravity, whose intensity is the boat's weight $F_G = mg$.

the liquid's thrust, which can be calculated by averaging the fulfilled hypotheses and then just as an Archimedean thrust, whose intensity F_A is equal to the weight of the liquid shifted by the immersed part of the hull. In this problem, the volume of the shifted liquid is V_{imm}:

The equilibrium is therefore written as:

$$F_A - F_G = 0 \qquad [2.135]$$

$$\rho V_{sub}\, g = mg \qquad [2.136]$$

h_0 can easily be deduced as:

$$\rho \frac{a\, h_0^2\, L}{2H} = m \qquad [2.137]$$

$$h_0 = \left(\frac{2\, H\, m}{\rho\, a\, L} \right)^{\frac{1}{2}} \qquad [2.138]$$

3) We can apply the precious reasoning with the new mass (m + M):

$$h = \left[\frac{2\, H\left(m + M\right)}{\rho\, a\, L} \right]^{\frac{1}{2}} \qquad [2.139]$$

The maximum value allowed for h before the boat sinks is H. We can immediately deduce the maximum value that M may take:

$$H = \left[\frac{2\, H\left(m + M_{Max}\right)}{\rho\, a\, L} \right]^{\frac{1}{2}} \qquad [2.140]$$

$$M_{Max} = \frac{\rho\, a\, L\, H}{2} - m \qquad [2.141]$$

4) Numerical applications:

4.1) The numerical value of h_0 is:

$$h_0 = 1{,}095\, m \qquad [2.142]$$

4.2) The numerical value of M_{Max} is:

$$M_{Max} = 2{,}9.10^{6}\ kg = 290\, t \qquad [2.143]$$

4.3) Only *2/3 of the boat's height can be immersed*. For this value of h, $h=\dfrac{2H}{3}$, the accepted value of M becomes:

$$\frac{2H}{3}=\left[\frac{2H\left(m+M_{Max}\right)}{\rho aL}\right]^{\frac{1}{2}} \qquad [2.144]$$

$$M_{Max}=\frac{\rho aL}{2}\frac{4H}{9}-m$$
$$M_{Max}=1,233.10^{5}\,kg \qquad [2.145]$$

If we remove the mass of the cargo (water, provisions and weapons), the acceptable mass of Vikings becomes:

$$M_{Vikings}=1,233.10^{5}-\left(500+1000+4000\right)=1,178.10^{5}=117,8\,t \qquad [2.146]$$

The number of Vikings is then:

$$n=\frac{1,178.10}{90}=1309 \qquad [2.147]$$

Such a number is sterically and, above all, sociologically difficult to conceive. For a bridge surface of 100 m², each warrior would only be granted a square of 27.6 cm each.

Moreover, in practice, we should think about allowing space for the horses that the Nordic warriors would not neglect to bring with them.

EXAMPLE 2.18 (Precautionary principle).–

In anticipation of an environmental catastrophe that we can always fear, in which Paris is submerged in water, let us input a safeguard strategy for the Eiffel Tower.

In order to save the Eiffel Tower, which has been registered as a UNESCO World Heritage site since 1991, we want to put it on floaters.

NOTE.– Currently, the Eiffel Tower's base is still at 33.5 m above sea level. Such a catastrophe is not expected to occur in the immediate future. Furthermore, we are not going to concern ourselves with the engineering or financing here.

Each of the tower's feet currently rest on a parallelepiped mass of concrete, with a horizontal section measuring 25 m^2 on each side and a depth of 15 m.

We will replace these bearings with parallelepipeds of the same horizontal section and a minimum height X, which we will determine. These bearings will be made out of expanded polystyrene. The Eiffel Tower's mass is $m = 10100t$.

The polystyrene's density is $\rho_B = 12 kg.m^{-1}$.

The water's density is $\rho = 1000 kg.m^{-1}$.

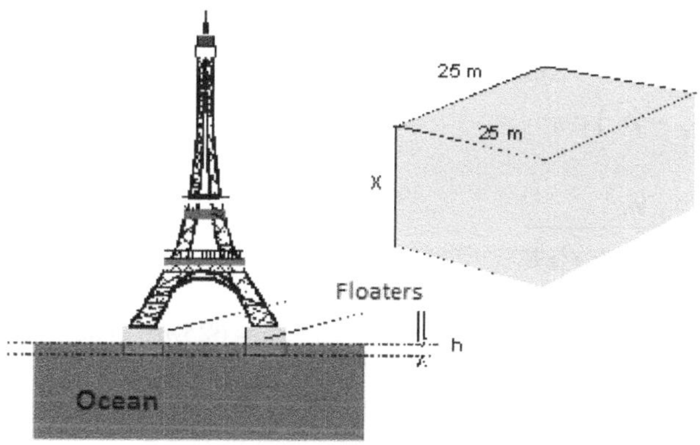

Figure 2.17. *The Eiffel Tower furnished for the downpour*

1) Calculate the minimum value of X so that the system floats and none of the Eiffel Tower's feet are in water.

2) Are you surprised by this result?

Solution:

1) What we have here is a solid statics problem. The Eiffel Tower must be at equilibrium under the action of the three forces, which are on a vertical axis:

Two downward forces; the force of gravity applied to the tower, four forces of gravity applied to each of the four floaters.

If each floater on the surface S is immersed at a height X, the Archimedean thrust is equal to the weight of the shifted fluid, or a volume SX, per floater, will be:

$$P_A = \rho SXg \qquad [2.148]$$

The weight of a floater is:

$$P_F = \rho_B SXg \qquad [2.149]$$

The equilibrium of the Eiffel Tower can therefore easily be written. All forces are colinear; the equilibrium therefore will take the form of an algebraic sum with the corresponding intensities:

Weight of the Eiffel Tower + Weight of the four floaters = Archimedean thrust on all floaters.

$$\begin{aligned} &mg + 4SX\rho_B\, g = 4SX\rho g \\ &4S\left(\rho_{water} - \rho_B\right) = m \end{aligned} \qquad [2.150]$$

$$X = \frac{m}{4S\left(\rho_{water} - \rho_B\right)} \qquad [2.151]$$

The air of each floater is equal to $S = 25*25 = 625\,m^2$. The minimum value of X then is:

$$\begin{aligned} &X = \frac{1,01.10^7}{4*625*(1000-12)} \\ &X = 4,089\,m \end{aligned} \qquad [2.152]$$

2) This may seem like a weak height. Nevertheless, it is worth observing that the pressure p_S exerted by the Eiffel Tower on the ground is much weaker than we would imagine at first sight.

$$\begin{aligned} &p_S = \frac{mg}{S} = \frac{10^7 * 9,81}{4*625} \\ &p_S = 3,9210^4\ Pa = 0,39\ bar \end{aligned} \qquad [2.153]$$

Furthermore, the floaters still represent a total volume of 10,000 m^3 of polystyrene.

2.3.3. *Calculating the thrust on a wall*

EXAMPLE 2.19 (Calculating the thrust on a flat vertical wall).–

A decanting tank is repaired by welding a rectangular plate with width l and height h that is perpendicular to the figure. The top of this piece is at the distance H from the free surface.

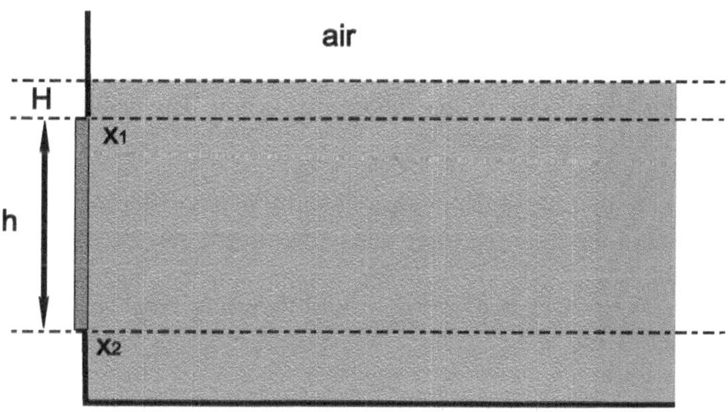

Figure 2.18. *The tank and its "patch"*

In order to estimate whether the repair will hold or not, we want to assess the force that the fluid with a density ρ exerts on the "patch".

1) Express this thrust $\vec{F}$.

2) Find the thrust's center. Is it in the center of the "patch"?

Solution:

1) In order to perform this calculation, we can observe that the pressure in the liquid is determined by the dimension z (altitude).

For the following calculation, it will be more practical to use a "depth" x, a distance on an axis directed downward from one point to the free surface.

In such a frame, pressure at a distance of x is determined by observing that in an incompressible fluid subject only to the forces of gravity, we increase the pressure $\rho g x$ by "descending" from "x":

$$p(x) = p_a + \rho g x \qquad [2.154]$$

The force that the fluid exerts on the door can therefore be calculated by cutting this door into horizontal strips (perpendicular to the figure) from the width l and infinitely small height dx. The pressure on each strip will be uniform and is given by the previous equation.

The fluid will apply a force $\vec{F}$ on each strip and air $dS = l\,dx$. The fluid will apply a thrust force (from right to left in the figure) $d\vec{F}$, with an intensity dF:

$$dF = p\,dx = \left(p_a + \rho g x\right) l\,dx \qquad [2.155]$$

On a vertical plate, all elementary forces are normal. The resultant intensities can be calculated from the algebraic sums.

One can observe that the ambient air will apply in a compensatory manner on the opposite face (left in the figure) a force due to a uniform pressure p_a.

The lower and higher ends of the door will therefore be at respective depths x_1 and x_2 given by:

$$\begin{aligned} x_1 &= h + H \\ x_2 &= H \end{aligned} \qquad [2.156]$$

The resultant force that the liquid applies will therefore be:

$$F = \int_{x_1}^{x_2} p\,dS = \int_{x_1}^{x_2} \left(p_a + \rho g x\right) l\,dx \qquad [2.157]$$

It is simple to integrate:

$$\begin{aligned} F &= l\left(p_a x + \rho g \frac{x^2}{2}\right)\Bigg|_{x_1}^{x_2} = p_a l\left(x_2 - x_1\right) + \rho g l \frac{x_2^2 - x_1^2}{2} \\ &= p_a l\left(x_2 - x_1\right) + \rho g l\left(x_2 - x_1\right)\frac{x_2 + x_1}{2} = p_a S + \rho g S \frac{x_2 + x_1}{2} \end{aligned} \qquad [2.158]$$

Finally, the force is:

$$F = S\left(p_a + \rho g x_M\right) = S p_M \qquad [2.159]$$

where:

$$x_M = \frac{x_2 + x_1}{2} = \frac{(H+h)+H}{2} = H + \frac{h}{2} \qquad [2.160]$$

We note that x_M is the depth of the door center and thus p_M is the ruling horizontal pressure that divides the door into two.

The total strain applied to the door will be the difference between this force F' due to the atmospheric pressure la p_a on the air S of the door:

$$F' = p_a\ S = p_a lh \qquad [2.161]$$

The resultant strain F_R is therefore equal to:

$$F_R = S(p_a + \rho g x_M) - p_a\ S = S\rho g x_M \qquad [2.162]$$

2) Nevertheless, the resultant strain of the difference between the pressure forces applied by the two sides of the wall will not be equivalent to that of the intensity force F_R applied to the center of the "patch". We need to find the application point or "thrust center" which would be the application point of the resultant whose strain would therefore be identical to the strains resulting from the pressure (especially from the point of view of shearing). The symmetries prove that this point will be in the center of a horizontal line with a depth x_C to be determined.

The thrust center C will be such that the moment of a force equivalent to the resultant forces applied on this point C will be equal to the resultant moment of the pressure forces. We will choose to calculate these moments against the upper side of the basin (with a depth of $x = 0$).

Therefore, we study the resultant moment of the pressure forces.

To simplify this, we will work from this point on each strip with a height dx, from the resultant antagonistic pressures on each side of the patch. We will also directly take into account the real effect born by the patch, from the beginning of the calculation.

Each elementary force $d\vec{F}$ (difference between the pressures applied on each side of the surface element) has an elementary module moment $d\vec{M}$:

$$dM = (p - p_a)x\,dS = (\rho g x)xl\,dx \qquad [2.163]$$

Again, the subtraction p_a assesses the forces of the liquid and atmospheric pressure applied on the two sides of the wall.

Furthermore, this resultant module must have the same module as the resultant moment placed in C :

$$\int dM = M_T = F_R x_C \qquad [2.164]$$

where F_R is the previously determined resultant thrust and x_C is the depth of the thrust center to be determined.

Let us calculate M_T:

$$M_T = \int_{x_1}^{x_2} (p - p_a) x \, dS = \int_{x_1}^{x_2} (\rho g x) x l \, dx \qquad [2.165]$$

where p is the pressure that the liquid applies onto the wall.

$$M_T = l \rho g \frac{x^3}{3} \bigg|_{x_1}^{x_2} = l \rho g \frac{x_2^3 - x_1^3}{3} \qquad [2.166]$$

x_C can thus be determined as:

$$\begin{aligned} & l \rho g \frac{x_2^3 - x_1^3}{3} = F_R \, x_C \\ & l \rho g \frac{x_2^3 - x_1^3}{3} = \rho g S \frac{x_2 + x_1}{2} x_C \\ & S = l(x_2 - x_1) \\ & l \rho g \frac{x_2^3 - x_1^3}{3} = \rho g l (x_2 - x_1) \frac{x_2 + x_1}{2} x_C \end{aligned} \qquad [2.167]$$

The general expression of x_C is:

$$x_C = \frac{2}{3} \frac{x_2^3 - x_1^3}{x_2^2 - x_1^2} \qquad [2.168]$$

In order to render this expression more legible, we will replace x_2 and x_1 by their function expressions of h and H. We have:

$$x_C = \frac{2}{3}\frac{(H+h)^3 - H^3}{(H+h)^2 - H^2} = \frac{2\left(h^3 + 3H^2 h + 3H h^2 \right)}{3\left(h^2 + 2H h \right)} \qquad [2.169]$$

Let us highlight a particular case. We can observe that if the upper part of the "patch" is at the surface level, meaning if $H = 0$, we find:

$$x_C = \frac{2}{3}\frac{x_2^3 - x_1^3}{x_2^2 - x_1^2} = \frac{2}{3}\frac{h^3}{h^2} = \frac{2}{3}h \qquad [2.170]$$

The thrust center in this case is centered on the patch at two-thirds lower than its height.

EXAMPLE 2.20 (Thrust on a porthole).–

A decantation tank is fitted with R-striped portholes, whose center is at a depth H below the free surface. The tank is filled with a liquid with a density ρ, which is assumed to be homogeneous.

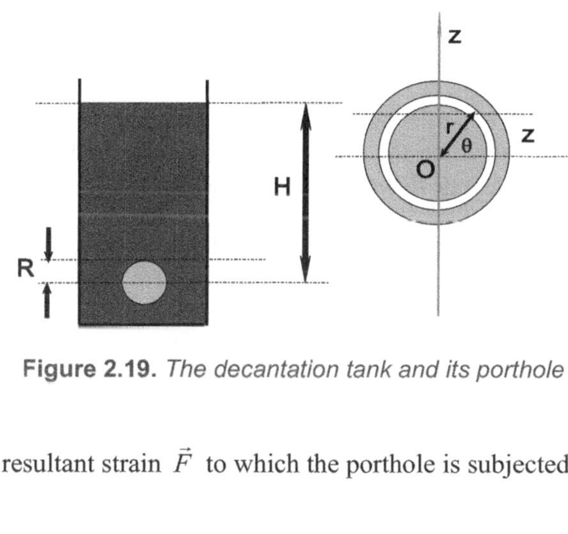

Figure 2.19. *The decantation tank and its porthole*

What is the resultant strain $\vec{F}$ to which the porthole is subjected?

We have:

$H = 10\,m$

$R = 60\,cm$

$\rho = 1024\,kg.m^{-3}$

Solution:

The problem involves fluid statics. The forces will be perpendicular to the porthole's surface. The resultant strain will be a normal force at each porthole. Using a vertical upward axis, we identify that the origin of the system will be at the center of the porthole.

The internal pressure on the porthole has a dimension z and is therefore given by:

$$p(z) = p_a + \rho g(H - z) \qquad [2.171]$$

We already assume that on each of the porthole's elementary surfaces, the atmospheric pressure will be applied onto the porthole's exterior in the opposite direction to the resultant internal pressure.

The resultant strain F is therefore expressed in the form:

$$F = \iint_{Porthole} (p(z) - p_a)dS = \iint_{Porthole} (p_a + \rho g(H - z) - p_a)dS = \iint_{Porthole} \rho g(H - z)dS \qquad [2.172]$$

Calculating F will result in a surface integral dealing with the contributions in $\rho g(H - z)$.

In order to perform this calculation, we identify a point on the porthole in a polar coordinate system centered on the porthole. The angles q will be measured starting from the same axis as the one that defines the dimensions.

In such a system, z is linked to the polar coordinates r and q by a simple relation:

$$z = r \sin\theta \qquad [2.173]$$

On a small field $dS = r\,dr\,d\theta$, the exceeding strain between the liquid's internal pressure and the external atmospheric pressure will be expressed by:

$$dF = \rho g(H - r\sin\theta)\ r\,dr\,d\theta \qquad [2.174]$$

The resultant strain will therefore be:

$$F = \int_0^{2\pi}\int_0^R \rho g(H - r\sin\theta)\ r\,dr\,d\theta \qquad [2.175]$$

This integral can easily be calculated as:

$$\int_0^{2\pi}(H - r\sin\theta)r\,dr\,d\theta = H\theta - r\cos\theta\Big|_0^{2\pi} = 2\pi H + 0$$

$$F = \int_0^R \rho g 2\pi Hrdr = 2\pi\rho g H\frac{R^2}{2} \qquad [2.176]$$

$$F = \pi R^2 \rho g H \qquad [2.177]$$

Thus, the resultant strain appears as the resultant strain per surface unit at the center of the porthole multiplied by the porthole's air.

Digitally, F is given by:

$$F = \pi R^2 \rho g H$$
$$F = 2,84.10^4\ N = 2895\ kf \qquad [2.178]$$

The strain approximately corresponds to 2.9 tons of force.

3

A Description of Flows

3.1. Introduction

The previous chapter was dedicated to immobile fluids. However, the most important aspect of fluid mechanics resides in the *dynamics of these fluids*; like in all branches of mechanical engineering, the movement is determined by the forces. This will be the focus of the following chapter.

However, before studying the determinants of a flow, we need to acquire a tool to precisely describe the movement of this flow. This description is the subject of *fluid kinetics*. Therefore, in this chapter, we will not concern ourselves with finding out *why* a flow is. Instead, we will be concerned with better informing the reader on *what it is*.

This chapter, which is dedicated to the description of a flow, will also be an opportunity to present the "first of the two physics principles" of fluid mechanics.

When writing "the two physics principles", we need to be aware of what *we are arbitrarily limiting to dynamics*. Placed where it should be within the realm of thermodynamics, "extended" fluid mechanics should integrate various other equations. In the first place, *the energy equation* is the expression adapted from the first principle of thermodynamics. While fluid is a mixture, it should also not be written as a *diffusion equation* by species. There are many types of transport present in the systems: viscosity, thermal conduction, possibly radiation and diffusion.

This work is dedicated to the first order of dynamics studies. Nevertheless, we will have to introduce the energy equation further on, which is dedicated to compressible fluids.

This chapter compiles:

a) At the tooling stage, the method for describing a flow. We will see that the method adopted differs from that which is familiar to the reader and used in mechanics of point power.

b) At the physics stage, how to write and apply the first physics principle, the principle of continuity. This principle, which is attached to the continuous nature of the matter implied, in fact expresses the conservation of matter. It is not necessary to express this principle with material elements, which are conserved because identified. In a continuous domain of fluid, the parts are not identified *a priori*.

3.2. The description of a fluid flow

3.2.1. *The Eulerian and Lagrangian description*

Describing a continuous matter's movements leads us to adopt a particular approach, which often misleads beginners and leads them to make errors of interpretation.

The first difficulty, and one that is quickly resolved, is when the fluid is continuous. We divide it into fluid particles, which can be as small as we wish. This then enables us to define an infinity of particles similar to material points. In simpler terms, our fluid becomes a group of infinite points constituting a continuous sphere. We note here that we overlook the discontinuous character of the fluid on the molecular scale. This discontinuity is indispensable for understanding and modeling a fluid's main properties, which include: compressibility, viscosity, thermal conduction and coefficient of diffusion. We should refer to any good physics textbook to understand this essential aspect. The molecular dimensions are much higher than the scale of fluid mechanics, by 10 orders of magnitude. Each scale has its own corresponding approach. For the fluid mechanics approach, matter is continuous. We recall that Newton himself did "not believe" in the possibility of an action from a distance. However, everything did in fact work "as if" it did. Physics is the child of modeling.

Nevertheless, we note that elementary particles are distinguished from the material point in that we cannot attribute a finite mass to it. We are therefore led to attribute it an equally elementary mass, expressed by using a "density". Let us take an example from everyday life.

If we want to provide information on the circulation of trains in a particular region, which we will liken to a flow of wagons, we can *a priori* adopt two strategies:

a) Acquire a publication which provides us with information on the situation of each train at every instant.

b) Ask the cows on the edges of the tracks to tell us which trains they have seen passing before them, at what time and at what velocity.

The first strategy refers to a timetable or, in a more modern context, the site of a railway company. This procedure appears to be the obvious choice when first analyzing it.

As outlined above, the second procedure seems rather torturous. However, it is fundamentally this one which will be adopted.

When transposed onto the scientific level of mechanical engineering, the first method refers to that which is familiar to everyone who has graduated from high school and having studied the mechanics of point power.

By following the "first" method proclaimed above, we can therefore describe a flow by giving the successive positions of each fluid particle at different instants. This particle is therefore identified by its position at the instant t = 0 (parameters a, b, c)

$$\begin{aligned} x &= x\,(t,a,b,c) \\ y &= y\,(t,a,b,c) \\ z &= z\,(t,a,b,c) \end{aligned} \qquad [3.1]$$

This is the point of view proposed by Lagrange, and for this reason, the first description will be called the Lagrangian description.

Distancing ourselves from the pleasant image of bovines aligned along the railway tracks, we can adopt a second point of view and, by placing ourselves in a vector position $\vec{r}(x,y,z)$ of a fixed frame, the vector is given for each instant. This is the velocity of the moving fluid particle at this point.

$$\vec{V} = \vec{V}\,(t,x,y,z),$$

or:

$$\begin{aligned} u &= u\,(t,x,y,z) \\ v &= v\,(t,x,y,z) \\ w &= w\,(t,x,y,z) \end{aligned} \qquad [3.2]$$

This second description mode will be called Eulerian.

Are these two descriptions complete and equivalent? The Lagrangian description gives us the velocity and (real) accelerations of the fluid particle being followed at each instant. Therefore, we also know what the velocity vector is for each fluid particle passing through at each given instant and point. In the same way, the data on the Eulerian velocity field theoretically allow us to calculate the successive positions of a fluid particle identified at a point and an instant; the trajectory is therefore re-established, as is the timed route.

There are two important remarks that need to be addressed at this point:

a) Unlike the habits in mechanics of point power, the triplet (x,y,z) is designated the position of the fixed point *where the observer is located*, rather than the moving matter.

b) The Eulerian description therefore gives the velocity of different particles, moving to a fixed point. In the Eulerian description, $\dfrac{\partial \vec{V}}{\partial t}$ *is in no way an acceleration*, because $\vec{V}$ relates to two *different* particles moving to the *same point* at *two different times* t and $t + dt$.

The physics here is Lagrangian, especially the dynamics principle, for which the acceleration is the second derivative with respect to the time of the Lagrangian vector position.

In order to deduce the real (Lagrangian) acceleration from the Eulerian velocity, we need to write the expression in a particular way. This will be the particular derivative or the Lagrangian derivative

$$\frac{d\vec{V}}{dt} = \frac{\partial \vec{V}}{\partial t} + \vec{V}.grad\,(\vec{V}) \qquad [3.3]$$

We recall its expression in Cartesian coordinates, in a projection on the axis i:

$$\frac{du_i}{dt} = \frac{\partial u_i}{\partial t} + u_j . \frac{\partial u_i}{\partial x_j} \qquad [3.4]$$

The partial derivatives $\frac{\partial \vec{V}}{\partial t}$ must be distinguished from the total velocity derivative $\frac{d\vec{V}}{dt}$, which constitutes the fluid particle's "real" acceleration (or the Lagrangian acceleration). We notice that, in a stagnant (or permanent) flow, $\frac{\partial \vec{V}}{\partial t}$ is zero, while $\frac{d\vec{V}}{dt}$ is generally not.

The description chosen will be the Eulerian description. A very good reason for this lies in the fact that fluid mechanics is in large part an experimental science and the observer is fixed against the laboratory, as well as the wind tunnel in which the flow is generated. The measurements are still Eulerian and the comparison with the model will be made on this basis.

3.2.2. *Kinematic elements*

The raw data on a Eulerian velocity vector are not sufficient to give an image of a flow. Furthermore, different demonstrations call upon accessory kinematic elements such as particle lines or tubes.

It will be necessary to distinguish the steady or permanent flows from the unsteady flows, which are more difficult to access when modeling.

Finally, there are some flows with a particularly simple structure which the reader needs to know. Furthermore, they constitute some very valuable information for resolving many questions.

a) We define different lines with the flow's velocity field:

The pathlines, constituted by points that are successively occupied by a given particle.

The streamlines, tangents at each of their points to the particle's velocity that is passing through at a given instant.

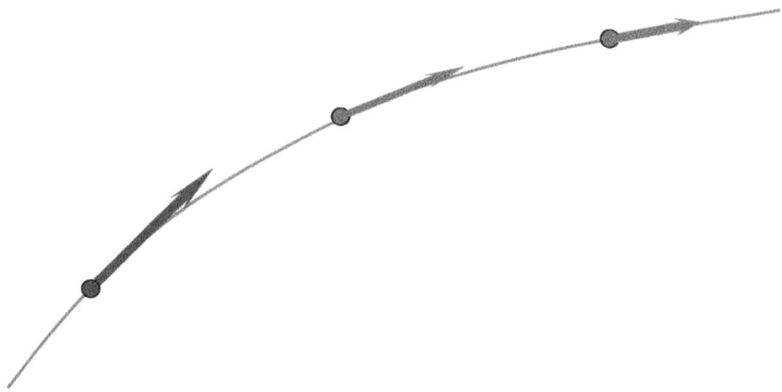

Figure 3.1. *Construction of a streamlines at an instant t. The arrows, like those in the following figures, represent, a fluid particle's velocity vector, which is individualized by its color*

The streaklines, are used less and are comprised of fluid particles at a given instant which have succeeded each other in the same point.

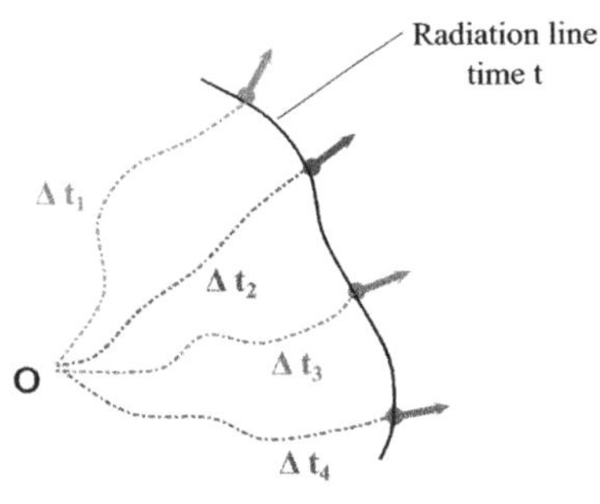

Figure 3.2. *Definition of streaklines. All of the particles located on the radiation line at the time dS_1 have moved to the "source" point O at previous instants t – Δ t, Δ t generally varies with the particle*

b) Starting from these lines, we construct particular surfaces or tubes:

Let us consider an imaginary finite surface S placed in a flow. This surface, which is not necessarily flat, possesses a contour, which is a closed line. We call this type of surface an "open surface", as opposed to a "closed surface", which delineates and defines a "volume". We construct a streamtube based on the group of current lines which move from the contour point of S.

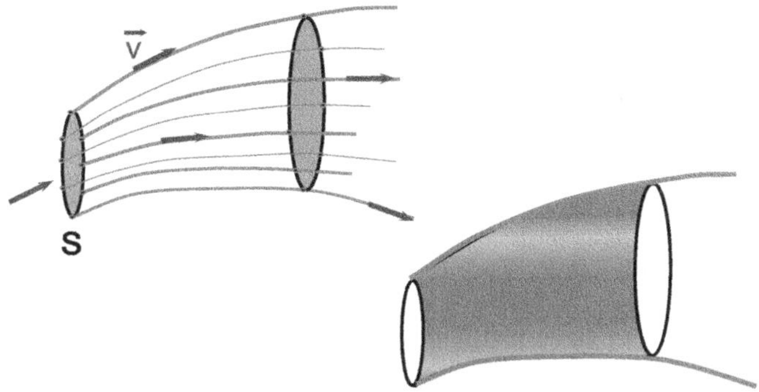

Figure 3.3. *Construction of a streamtube. This tube is a continuous lateral surface. The flow that crosses each of its sections is not generally uniform*

If the surface becomes an infinitely small surface *dS*, which we will later call an "elementary" surface, the current tube takes the name of the elementary streamtube.

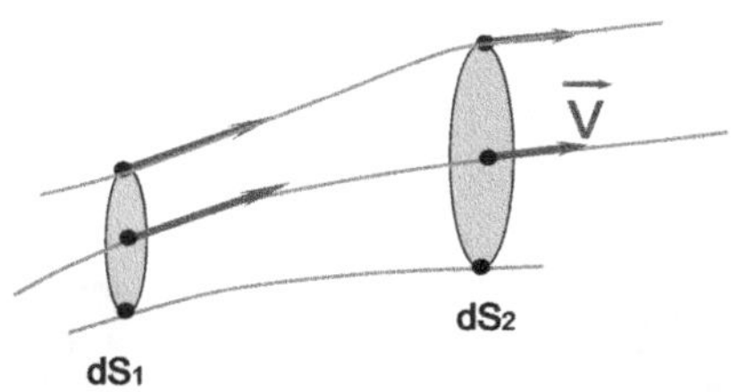

Figure 3.4. *Construction of an elementary streamtube. The flows that cross dS_1 and dS_2 are locally uniform*

c) Notion of steadiness of flow:

We distinguish the steady flows from the unsteady (or permanent) flows.

Using a Eulerian description, in *unsteady* flows, the velocity varies with time. The fluid particles which succeed each other in a fixed point before the observer have different velocity vectors. Each fluid particle passing a fixed point has had, and therefore will have from this point onwards, a different path and timed route.

In *permanent or steady* flows, each fluid particle passing from a point has the same velocity vector as the previous one.

The term "permanent flow" is rather commonly used among fluid mechanical engineers. We often prefer to use the term "stagnant" here, which is opposed to the notion of non-stagnation. The majority of the examples in this book are also dedicated to stagnant flows.

Time no longer appears in the Eulerian description of a stagnant flow (which does not stop it from having a Lagrangian acceleration $\frac{d\vec{V}}{dt}$). Particles that pass by all fixed points (x, y, z) have had, and will have from this point onwards, the same timed route.

Permanent flows will constitute the majority of the flows studied in the problems of this book.

In permanent flows, the current lines, trajectory and radiation lines are fixed in time and geometrically confounded. However, we must not lose sight of the fact that these lines remain totally different in nature.

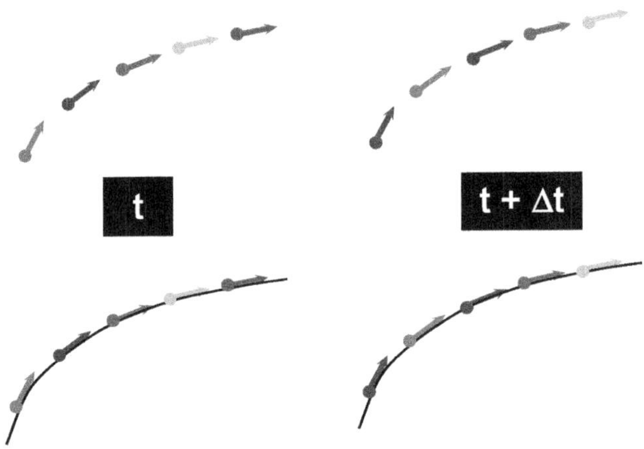

Figure 3.5. *In steady (or permanent) regimes, the network of streamlines is invariable. It is confounded with that of pathlines. On the left-hand side, there is a velocity field and flow line at time t ; on the right-hand side, there is a velocity field and flow line at time t + dt*

In an unsteady flow, the streamlines, pathlines and streaklines are variable in time and are geometrically distinct from this fact.

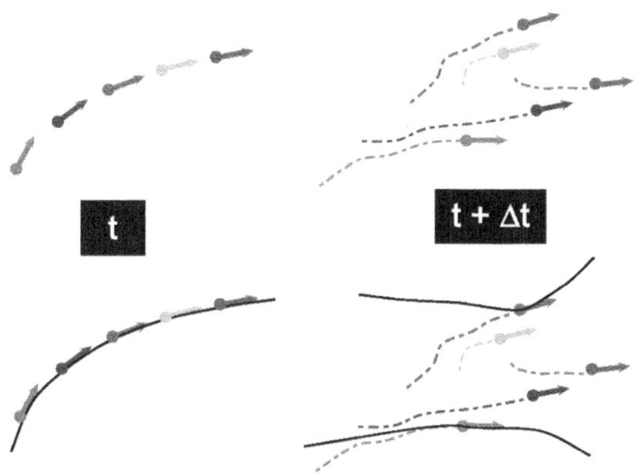

Figure 3.6. *In an unsteady flow, streamlines vary from one instant to another. On the left-hand side, there is a velocity field and streamline at time t; on the right-hand side, there is a velocity field and streamline at the time t + dt*

Furthermore, we note that, in unsteady flows, a streamtube only has a direction at a given instant. The streamlines of a steady flow are actually mobile in time and the streamtube changes geometrically according to the instant in question.

This is not the case for steady flows. The construction of a streamtube (elementary or not), when based on the fixed surfaces S or dS, becomes particularly useful, as tubes become permanent in time.

d) In this context, various geometrical flows are remarkable. In a Eulerian description, it is as follows:

Parallel flows or velocity vectors are all parallel, *but be careful – they are not equipollent*, their standards can vary from one point to another.

Let us look at a particular case of a *uniform flow*, where all velocity vectors are equipollent. From a physics point of view, this is a flow "in a block" (all fluid particles move at the same speed, just like non-deformable solids).

Parallel and uniform flows will play a major role in processing the examples dealt with in the dynamics of perfect fluids, where flows are considered to be related to a mobile one. In this configuration, the flow "far" ahead of the mobile one is a purely uniform flow.

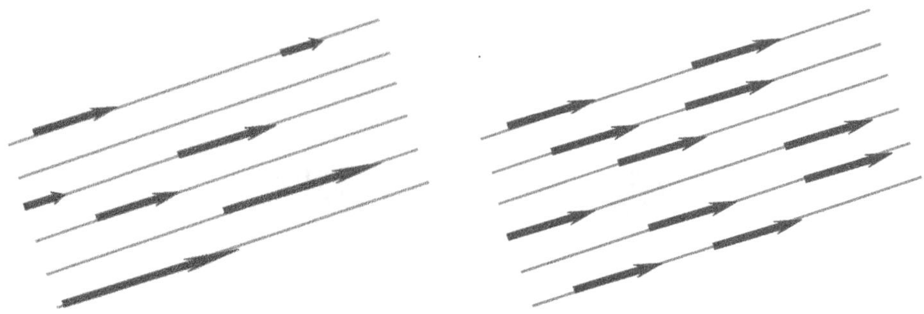

Figure 3.7. *The uniform flow (right) is a particular case of a parallel flow (left). On the right-hand side, all the velocity vectors are equipollent, which is not verified on the left-hand side*

In practice, a flow is rarely parallel or uniform everywhere, although they may be in certain geographically limited areas. In such a case, we would have a *locally parallel* or *locally uniform* flow. We will see that this valuable property is leveraged when resolving many problems.

Henceforth, we note the case of elementary streamtubes. As a matter of fact, all fluid particles crossing an elementary surface dS at a given instant will have the same velocity vector. The flow that crosses dS is what is called locally uniform.

NOTE.– If the elementary streamtube has, as is often the case, a variable section along a current line, then we will see that the principle of continuity will imply a speed variation from one dS to another.

3.2. A first principle of physics: the principle of continuity

3.2.1. *The principle of continuity*

This constitutes the first of the two principles to be written in fluid mechanics.

3.2.1.1. *Notions of flow rate and calculations*

In order to outline and understand this principle, we will need the notion of flow rate.

When an open surface S is placed in a flow, a certain fluid "quantity" crosses this surface by time unit. We have already noted that, on a finite surface, this crossing will follow in the places of S and will be performed in one "direction" or another. The flow rate therefore constitutes the balance sheet of these crossings in opposite directions. This distinction is no longer imposed for elementary surfaces, crossed by a locally uniform flow with a unique velocity $\vec{V}$.

The certain fluid "quantity" that crosses a surface by time unit can be expressed either in volume terms or mass terms.

A volumetric flow rate through a surface S will be defined thus as the volume passed by time unit from one side of S to the other.

A mass flow rate through a surface S will be defined thus as the fluid mass passed by time unit from one side of S to the other.

Generally, for unsteady compressible flows, the flow rate must be understood as instantaneous and be defined and calculated on an elementary time dt. Thus, we have the volumetric flow rate and the mass flow rate, respectively:

$$Q_V = \frac{dV_{OL}}{dt} \qquad [3.5]$$

$$Q_m = \frac{dM}{dt} \qquad [3.6]$$

These definitions also hold true for flow rates crossing an elementary surface. These flow rates will also be elementary:

$$dq_V = \frac{d^2V_{OL}}{dt} \qquad [3.7a]$$

$$dq_m = \frac{d^2m}{dt} \qquad [3.7b]$$

When passing through a finite surface, the density can vary from one point to another. When passing through an elementary surface, the density ρ will remain uniform, even for compressible fluids.

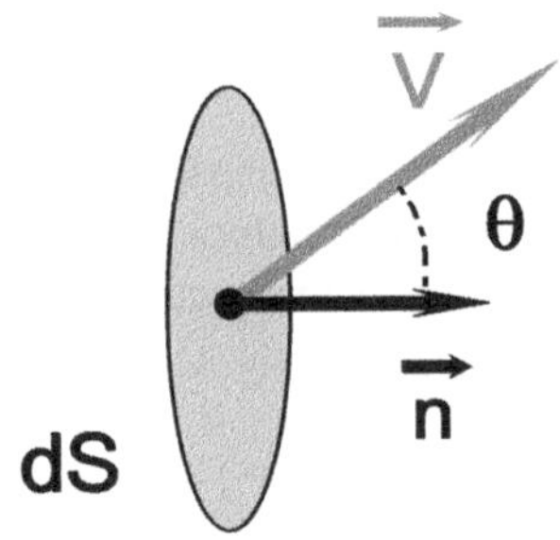

Figure 3.8. *Mass and volumetric flow rate calculation components when passing through an elementary surface*

It is demonstrated that the elementary volumetric flow rate dq_V, when passing through an elementary surface whose normal $\vec{n}$ makes an angle θ with the velocity vector $\vec{V}$ of the locally uniform flow which crosses it, is written as:

$$dq_V = V\cos\left(\vec{n},\vec{V}\right)dS = V\,dS\cos\theta = \vec{V}.\,\vec{n}\,dS \qquad [3.8]$$

The last expression is explained by noticing that the norm of the unitary vector $\vec{n}$ is $n = 1$.

Therefore:

$$V\,dS\cos\theta = \left|\vec{V}\right|.\left|\vec{n}\right|\cos\theta = \vec{V}.\,\vec{n} \qquad [3.9]$$

This flow rate for a normal flow on dS becomes:

$$dq_V = V\,dS \qquad [3.10]$$

The previous observation immediately leads to the expression of the mass flow rate:

$$dq_m = \rho V\,dS\cos\theta = \rho\vec{V}.\,\vec{n}\,dS \qquad [3.11]$$

which, for a normal flow on dS, becomes:

$$dq_m = \rho V\,dS \qquad [3.12]$$

These expressions are fundamental and will be used systematically in the examples of this book.

Regardless of whether via an open or closed surface, the volumetric and mass flow rates are written in a general way as:

$$Q_V = \iint_S dq_V = \iint_S \vec{V}.\vec{n}\, dS = \iint_S V \cos\theta\, dS \qquad [3.13]$$

$$Q_m = \iint_S dq_m = \iint_S \rho\vec{V}.\vec{n}\, dS = \iint_S \rho V \cos\theta\, dS \qquad [3.14]$$

Calculating these flow rates will be more or less complex depending on the flow.

We note that the sign of the flow rate found will depend on the direction of the normal one. The flow rate found will generally be positive if more fluid passes in the same direction as the normal one than in the opposing direction of this normal flow rate.

This observation becomes all the more significant in the case of closed surfaces. Indeed, the integral formulas of vectorial geometry are only valid if the normal standards are turning *outwards* from this closed surface. What results is a flow rate that is calculated to be positive for the fluid emerging from a closed surface. We must be aware of this particularity when writing the principle of continuity through a closed surface.

There is significant and particular case to be retained here. When a locally uniform flow crosses through a finite surface S:

$$Q_V = \iint_S V \cos\theta\, dS = VS \cos\theta \qquad [3.15]$$

If, moreover, the fluid is incompressible:

$$Q_m = \iint_S \rho V \cos\theta\, dS = \rho V\, S \cos\theta \qquad [3.16]$$

This is particularly the case when calculating a volumetric and mass flow rate in the right section of a tube travelled by a perfect fluid. In this case, the fluid is normal in the section and:

$$Q_V = VS \qquad [3.17]$$

$$Q_m = \rho VS \qquad [3.18]$$

We finally notice that the lateral wall of a streamtube is constituted by flowlines. All elementary surfaces of this lateral wall therefore have a normal vector that is perpendicular to the current lines, and therefore to the velocities. They are not crossed by any flow rate: no fluid can enter or emerge from the lateral walls of a streamtube. This obviously applies to the particular tubes that are elementary streamtubes. *The mass flow rate of a fluid traveling inside a streamtube is thus conserved.*

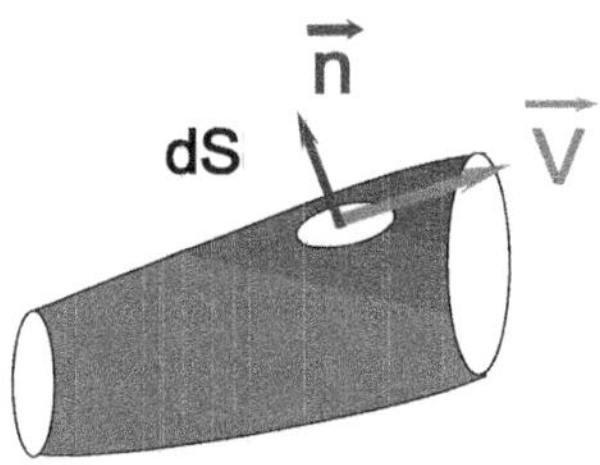

Figure 3.9. *The fluid cannot cross the walls of a streamtube*

3.2.1.2. *Principle of continuity*

Depending on the "level" of the problem to be solved, this principle is written either more or less simply and either in an analytic or integrated way.

In any case, the principle is based on what Lavoisier expressed in 1789, which the following phrase made apocryphal: "nothing is lost, nothing is created".

Anaxagoras already had an intuition of this principle when he said "nothing is born or perishes, but things which already exist combine together and then separate again".

We will apply these principles here in the case of a reputedly continuous fluid. We have already noticed that this "continuity", on the other hand, results from a mobilizing approach.

The principle of continuity is expressed in terms of a mass balance on a closed surface: the variation of mass, contained in a domain D limited by a surface S that is fixed against the Eulerian reference frame, is equal to the difference between the mass flow rates that enter and emerge from this domain, crossing S.

When writing this balance sheet in Cartesian coordinates for a small area of edges dx, dy, dz, we obtain the principle of continuity, in the form of a local differential equation, a "point equation".

A compressible (ρ variable) and non-stagnant flow in this point equation or continuity equation is generally written as:

$$\frac{\partial \rho}{\partial t} + \frac{\partial \rho u}{\partial x} + \frac{\partial \rho v}{\partial y} + \frac{\partial \rho w}{\partial z} = 0 \qquad [3.19]$$

Using the vectorial analysis concepts, we introduce the divergence of the speed vector, which is defined in Cartesian terms by:

$$div\vec{V} = \frac{\partial u}{\partial x} + \frac{\partial v}{\partial y} + \frac{\partial w}{\partial z} = 0 \qquad [3.20]$$

We can therefore rewrite the continuity equation in the following form:

$$\frac{\partial \rho}{\partial t} + div\left(\rho\vec{V}\right) = 0 \qquad [3.21]$$

The use of this form is that it can be applied to all the reference frames of coordinates, and thus enables us to rewrite the equation in cylindrical or spherical terms, for example.

We often find a flat, stagnant or incompressible flow, for which the point equation for continuity is reduced to:

$$\frac{\partial u}{\partial x} + \frac{\partial v}{\partial y} = 0 \qquad [3.22]$$

or: $$div\vec{V} = 0 \qquad [3.23]$$

We can show that $div\vec{V} = 0$ is a *necessary and sufficient condition* for a fluid to be incompressible. Some authors further pose that a zero divergence is the definition of incompressibility. This usage strikes us as too abstract for the context of this publication.

3.3. Notions and recalls on potential flows

3.3.1. *Definition*

The study of potential flows is only of interest for the chapter on perfect fluids. It is often a prerequisite to studying a viscous fluid's interactions with a wall. Further on (in Chapter 6, when dealing with boundary layers), we will see that the following solution strategy is then adopted:

a) Flows generated by introducing a solid body into the uniform flow by a theory, assumed to be a perfect fluid. This flow is called "potential". Thus, a "potential" flow is therefore calculated by assuming a perfect slide from the fluid to the wall.

b) We take account of the viscous fluids' "non-sliding to the wall" by an always "geometrically thin" junction between the potential flow and the wall. The potential flow, called "the far flow", is therefore "perturbed" by a boundary layer, which constitutes an "asymptotic junction" between the perfect flow and the wall.

3.3.2. *Determination*

For a *flat, stagnant and incompressible flow*, the potential flow's calculation goes through the solution for a Laplace equation.

Point equations for incompressible and stagnant flows lead to two representative function fields being introduced from the flow's kinematics: potential functions $\Phi(x,y)$ and function of the current $\Psi(x,y)$.

These definitions are particularly useful when the fluids are also perfect. They slide perfectly onto the solid walls and we can write that the velocity is tangential to the wall:

$$div\vec{V} = \frac{\partial u}{\partial x} + \frac{\partial v}{\partial y} = 0 \qquad [3.24]$$

The potential function Φ is introduced by writing that $\vec{V}$ is the gradient of a potential function:

$$\vec{V} = gr\vec{a}d\,\Phi \qquad [3.25a]$$

$$u = \frac{\partial \Phi}{\partial x}; \quad v = \frac{\partial \Phi}{\partial y} \qquad [3.25b]$$

We notice that unlike the derivation of a potential force, here $\vec{V} = + \vec{grad}\,\Phi$.

In the case of the forces, the gradient is preceded by a lesser one, whose script is motivated by the formulation of energy conservation $E_C + \phi = Cte$.

The function of current Ψ is defined by:

$$u = \frac{\partial \Psi}{\partial y}; \quad v = -\frac{\partial \Psi}{\partial x} \qquad [3.26]$$

Through the definition of Ψ, the continuity equation is automatically verified.

On the other hand, by introducing the potential function into an incompressible fluid's continuity equation, it is shown that Φ satisfies Laplace's equation:

$$div\vec{V} = \frac{\partial u}{\partial x} + \frac{\partial v}{\partial y} = \frac{\partial}{\partial x}\frac{\partial \Phi}{\partial x} + \frac{\partial}{\partial y}\frac{\partial \Phi}{\partial y} = 0 \qquad [3.27]$$

$$\Delta\Phi = \frac{\partial^2 \Phi}{\partial x} + \frac{\partial^2 \Phi}{\partial y} = 0 \qquad [3.28]$$

Two networks of curves can be defined thus: the network of equipotentials and the network of curves $\Psi = Cte$.

We also notice that a direct calculation of the components of a rotational velocity shows that they are at zero. Here, we find a classic property of the Curl of the gradient.

The gradient vector is normal to the equipotential, and hence, the equipotentials $\Phi = Cte$ are normal to the velocity vectors $\vec{V}$. The equipotentials are therefore at each of their normal points in the network of current lines.

By following an analogous reasoning, we can show that the curves $\Psi = Cte$ are current lines.

To demonstrate this, we only need to calculate the scalar product of $\vec{V}$ and the gradient of Ψ, which is normal to the lines $\Psi = Cte$:

$$\vec{V}.\vec{grad}\,\Psi = \frac{\partial \Psi}{\partial y}\frac{\partial \Psi}{\partial x} - \frac{\partial \Psi}{\partial x}\frac{\partial \Psi}{\partial y} = 0 \qquad [3.29]$$

$\vec{V}$ is normal to $\overrightarrow{grad}\,\Psi$, therefore tangential to the lines aux $\Psi = Cte$, which are definitely current lines.

The network of the curves $\Phi = Cte$ and $\Psi = Cte$ is therefore orthogonal.

Historically, we would attempt to define equipotentials using a given solid body (which itself constitutes an equipotential by an electrical analogy); therefore, we would use a real-time tank. Today, using digital techniques to calculate Laplace's equation is the method that prevails. This question will be broached in the part of this book dedicated to digital approaches.

3.3.3. *Determining streamlines*

We often attempt to determine current lines using a given bi-dimensional flow in a Eulerian form. To do this, we can find the current lines by writing that, on the current line:

$$\Psi = Cte \tag{3.30}$$

There is another elegant (and equivalent) way of doing this, which consists of noticing that an elementary vector, which is tangential to the current lines $d\vec{l}\,(dx,dy)$ is co-linear to the velocity $\vec{V}(u,v)$. Their following respective components x and y are therefore proportional:

$$\frac{dx}{u} = \frac{dy}{v} \tag{3.31a}$$

Thus, we obtain an equation with a differential form, which can also be put in the more practical form:

$$vdx = udy \tag{3.31b}$$

which can be processed by finding the function $y = y(x)$.

3.3.4. *Curl of the velocity*

Although this veers away from the main theme of the book, we will mention the use brought to the rotational velocity $\overrightarrow{rot}\,\vec{V}$. As for certain flows in perfect fluids, the velocity is derived from a potential and this Curl of the velocity is a zero vector.

We will see a few examples of them in the case of a number with limited problems. In other cases, particularly for real fluids, the appearance of "velocity gradients", which generate viscosity fluids, leads to the fluid particles having rotational effects. The Curl of the velocity $\vec{rot}\,\vec{V}$ therefore becomes the double of the whirlwind vector $\vec{\Omega}$ that appears as the rotation vector of the fluid particle. We refer the reader to a more specialized publication to read more on this subject.

In support of the following example, remember that, in Cartesian coordinates, the Curl of the velocity is written as:

$$\left(\vec{rot}\,\vec{V}\right)_x = \frac{\partial w}{\partial y} - \frac{\partial v}{\partial z} \qquad [3.32a]$$

$$\left(\vec{rot}\,\vec{V}\right)_y = \frac{\partial u}{\partial z} - \frac{\partial w}{\partial x} \qquad [3.32b]$$

$$\left(\vec{rot}\,\vec{V}\right)_z = \frac{\partial v}{\partial x} - \frac{\partial u}{\partial y} \qquad [3.32c]$$

Stokes' theorem establishes a useful relation between velocity circulation $\vec{V}$ on a closed contour C and the rotational flux of this velocity $\vec{rot}\,\vec{V}$ through any supported surface S on this contour C:

$$\int_C \vec{V}.d\vec{l} = \iint_S \vec{rot}\,\vec{V}.\vec{n}\,dS \qquad [3.33]$$

This relation will be useful when studying the free vortex (Example 3.5).

Finally, here we notice a relation that will be developed in the Appendix and which links the Lagrangian acceleration to this curl of the velocity:

$$\frac{d\vec{V}}{dt} = \frac{\partial \vec{V}}{\partial t} + \vec{grad}\frac{V^2}{2} - \vec{V} \wedge \vec{rot}\,\vec{V} \qquad [3.34]$$

3.4. Example of kinematic calculations

General remarks: in simple problems, we use the principle of continuity by writing it in a more global form. This is often the case for stagnant flows. In such a flow, the mass contained in a domain does not vary. *Therefore, we determine that in the system constituting the problem, an area seems to be pertinent and we write that the sum of "entering" flow rates is equal to the sum of "emerging" flow rates.*

A particular case resides in a section of a current tube (or net). We then equalize the two flow rates crossing the surfaces that are leaning against the current tube, because no flow rate is crossing the tube in question. In the same perspective, we will also find sections of solid tube.

We are often then met with the problem of a flowing fluid's incompressibility. In this case, we verify whether the divergence is canceled out or not.

Another standard problem lies in determining the current lines for a given flow in a Eulerian description. We would then mainly resort to the relations [3.31a] and [3.31b]

EXAMPLE 3.1.–

A flat flow is described in Eulerian form by a speed vector whose components u and v, respectively, following the axes Ox and Oy on the flat level, are given by:

$$u = a \quad [3.35]$$

$$v = a u_0 x \quad [3.36]$$

where a and u_0 are two constants.

1) Is it possible to observe such a flow with an incompressible fluid?

2) Give the expression of its streamlines. What are the curves that make up this network of streamlines?

3) Resume questions (1) and (2) in the case of a velocity vector given by:

$$u = axy \quad [3.37]$$

$$v = a u_0 xy \quad [3.38]$$

Solution:

1) We calculate the divergence from the velocity:

$$div\,\vec{V} = \frac{\partial u}{\partial x} + \frac{\partial v}{\partial y} \quad [3.39]$$

$$div\,\vec{V} = 0 + 0 \quad [3.40]$$

The divergence from the velocity is zero. The flow can be incompressible.

2) We use the relations [3.31]:

$$\frac{dx}{u} = \frac{dy}{v}; \ vdx = udy \tag{3.31}$$

$$a\,u_0\,x\,dx = a\,dy \tag{3.41a}$$

$$u_0\,x\,dx = dy \tag{3.41b}$$

$$u_0 d\frac{x^2}{2}x = dy \tag{3.42}$$

$$y = \frac{x^2}{2} + Cte \tag{3.43}$$

which is the equation of a network of parabolas whose peaks are on the Oy.

3) By picking up the same reasoning stages and calculations as in (1) and (2):

$$u = axy \tag{3.44}$$

$$v = au_0\,xy \tag{3.45}$$

We ask ourselves about the incompressibility:

$$div\,\vec{V} = \frac{\partial u}{\partial x} + \frac{\partial v}{\partial y} \tag{3.46}$$

$$div\,\vec{V} = ay + a\,u_0\,x = a\left(y + u_0\,x\right) \tag{3.47}$$

The divergence from the velocity is only zero in some singular points. The fluid therefore cannot be incompressible.

We search for the equation of the current lines (equation [3.31]):

$$\frac{dx}{u} = \frac{dy}{v} \tag{3.48}$$

The alternative form seems more practical:

$$vdx = udy \qquad [3.49]$$

$$a u_0 \, xy \, dx = axy \; dy \qquad [3.50a]$$

$$u_0 \, dx = dy \qquad [3.50b]$$

or by integrating:

$$y = u_0 \, x + Cte \qquad [3.51]$$

The current lines constitute a network of straight parallel lines with a slope u_0.

EXAMPLE 3.2.–

We give a flat flow in a Eulerian description by its velocity's two components, u and v:

$$u = \frac{x^n}{a} \qquad [3.52]$$

$$v = \frac{y^m}{b} \qquad [3.53]$$

where a, b, n and m are constants.

1) Is this flow steady or unsteady?

2) Is the fluid compressible or incompressible?

3) Find the equation for such a flow's current lines. From this, deduct the equation for the trajectories of its fluid particles.

Solution:

1) The expression of the velocity components does not contain time. The flow is stagnant (or permanent).

2) If the fluid is incompressible, the velocity components for this steady, flat and bi-dimensional fluid must verify:

$$div\vec{V} = 0 \qquad [3.54]$$

or:

$$\frac{\partial u}{\partial x}+\frac{\partial v}{\partial y}=0 \qquad [3.55]$$

$$\frac{\partial}{\partial x}\frac{x^n}{a}+\frac{\partial}{\partial y}\frac{y^m}{b}=\frac{nx^{n+1}}{a}+\frac{my^{m+1}}{b} \qquad [3.56]$$

This expression is not zero *a priori* in the whole flow. The fluid is therefore compressible.

3) The streamlines will be determined based on:

$$v\,dx = u\,dy \qquad [3.57]$$

Here:

$$\frac{y^m}{b}dx=\frac{x^n}{a}dy \qquad [3.58]$$

which integrates into:

$$\frac{a\,dx}{x^n}=\frac{b\,dy}{y^m}$$
$$\frac{by^{1-m}}{1-m}=\frac{ax^{1-n}}{1-n}+C \qquad [3.59]$$

From this, the below results:

$$y=\left[\frac{a(1-m)}{b(1-n)}x^{1-n}+C\right]^{\frac{1}{1-m}} \qquad [3.60]$$

The flow is permanent, and therefore, the current lines are also the trajectories of the fluid particles.

EXAMPLE 3.3.–

We imagine a flat and incompressible flow whose two components in a Cartesian reference frame *Oxy*, respectively $u(x,y)$ and $v(x,y)$ are in the following form:

$$u = ax^2y \qquad [3.61]$$

$$v = axy^2 \qquad [3.62]$$

where a and b are two constants:

1) Which condition must a and b fulfill in order for the *incompressible* flow to be possible?

2) Under this condition, give the equation of this flow's current lines. What type of curves are these?

3) Give the fluid's Lagrangian acceleration components.

Solution:

1) Neither the expression of u, nor that of v explicitly contain the time. Therefore, the flow is stagnant.

An incompressible flow has zero divergence. We should therefore have:

$$div\,\vec{V}=0;\ \frac{\partial u}{\partial x}+\frac{\partial v}{\partial y}=0 \qquad [3.63]$$

For this flow:

$$div\,\vec{V}=2axy+2bxy \qquad [3.64]$$

which will be zero on the condition that:

$$a=-b \qquad [3.65]$$

The velocity components are therefore in the form:

$$u=ax^2y \qquad [3.66]$$

$$v=-axy^2 \qquad [3.67]$$

2) The equation for current lines is deducted from:

$$v\,dx=u\,dy \qquad [3.68]$$

$$-axy^2\,dx=ax^2y\,dy \qquad [3.69]$$

or even:

$$-\frac{dx}{x}=\frac{dy}{y} \qquad [3.70]$$

which integrates into:

$$Ln\, y = -Ln\, x + Ln\, C \qquad [3.71a]$$

$$y = \frac{C}{x} \qquad [3.71b]$$

The current lines make up a network of hyperboles, axes (Ox, Oy) and equations:

$$xy = C \qquad [3.72]$$

3) The Lagrangian acceleration $\vec{\Gamma}$ is deduced from the definition:

$$\Gamma_x = \frac{du}{dt} = u\frac{\partial u}{\partial x} + v\frac{\partial u}{\partial y} \qquad [3.73]$$

$$\Gamma_y = \frac{dv}{dt} = u\frac{\partial v}{\partial x} + v\frac{\partial v}{\partial y} \qquad [3.74]$$

$$\Gamma_x = ax^2y\frac{\partial\, ax^2y}{\partial\, x} - axy^2\frac{\partial\, ax^2y}{\partial\, y} = 2a^2x^3y^2 - a^2x^3y^2 = a^2x^3y^2 \qquad [3.75]$$

$$\Gamma_y = -ax^2y\frac{\partial\, axy^2}{\partial\, x} + axy^2\frac{\partial\, axy^2}{\partial\, y} = -a^2x^2y^3 + 2a^2x^2y^3 = a^2x^2y^3 \qquad [3.76]$$

We notice that on the bisectors of the axis system ($x = y$ or $x = -y$), the acceleration is carried by the corresponding bisector:

$$\Gamma_x = a^2x^5 = a^2y^5 = \Gamma_y \qquad [3.77]$$

or even $\Gamma_x = -a^2x^5 = -a^2y^5 = \Gamma_y$ [3.78]

EXAMPLE 3.4 (Solid vortex).–

In a Cartesian reference frame, a flat flow is defined by the following velocity field:

$$u = -Ay \qquad [3.79]$$

$$v = Ax \qquad [3.80]$$

where A is a constant.

1) Determine the current lines and trajectories of this flow's fluid particles.

2) Show that this flow is incompressible.

3) Calculate the velocity module for each point $M(x,y)$ of this flow's flat line.

4) Calculate the acceleration vector field $\vec{\Gamma}=\dfrac{d\vec{V}}{dt}$. Find its module (or standard) Γ.

5) Calculate the rotational field $Curl\,\vec{V}$.

6) Were these results for questions (4) and (5) not foreseeable without calculating them?

Solution:

1) The flow is stagnant. The current lines and trajectories are networks of superimposed curves. We will use the relation expressing that an infinitesimal line that is tangential to the current line $d\vec{l}\,(dx,dy)$ is collinear to the velocity vector $\vec{V}(u,v)$ (by definition the current line):

$$\frac{dx}{u}=\frac{dy}{v} \qquad [3.81]$$

In practice, it is more useful to rewrite the expression in the form:

$$v\,dx = u\,dy \qquad [3.82]$$

For the components u and v given here:

$$Ax\,dx = -\,Ay\,dy \qquad [3.83]$$

which easily integrates into:

$$\frac{y^2}{2}=-\frac{x^2}{2}+Cte \qquad [3.84]$$

At this stage of reasoning, we could express the equation of the current line in the explicit form $y = \sqrt{x^2 + 2Cte}$, which is not very striking *a priori*.

A bit of intuition allows us to notice that $x^2 + y^2 = r^2$, where r is the distance from one point of a current line to the source of the coordinates (in other terms, the norm of the position vector $\vec{r}$ from one point in our frame)

We can therefore rewrite [3.84] in the form:

$$x^2 + y^2 = r^2 = C^2 \qquad [3.85a]$$

$$r = C \qquad [3.85b]$$

Thus, the geometry of the current lines becomes evident. The current lines are circles centered on the source of the coordinates.

2) If a fluid is incompressible, the flow will verify:

$$\begin{aligned} &div\vec{V} = 0 \\ &\frac{\partial u}{\partial x} + \frac{\partial v}{\partial y} = 0 \end{aligned} \qquad [3.86]$$

For this flow:

$$div\vec{V} = -\frac{\partial Ay}{\partial x} + \frac{\partial Ax}{\partial y} = 0 \qquad [3.87]$$

The fluid is therefore incompressible.

3) The V module of velocity is calculated based on u and v:

$$V^2 = u^2 + u^2 = A^2x^2 + A^2y^2 = A^2r^2 \qquad [3.88]$$

$$V = Ar \qquad [3.89]$$

On this level, we can make an image of this flow. Each point of the fluid has a circular trajectory, with a velocity that is proportional to the distance from the source. The fluid therefore behaves like a block that turns on its axis Oz. This flow is often named the "solid" vortex, as opposed to the "free" vortex, where the velocity module is in $V = \frac{1}{r}$ (see Example 3.5)

4) We recall the Cartesian expression of the rotational components:

$$Curl\,\vec{V} = \begin{matrix} \dfrac{\partial w}{\partial y} - \dfrac{\partial v}{\partial z} \\ \dfrac{\partial u}{\partial z} - \dfrac{\partial w}{\partial x} \\ \dfrac{\partial v}{\partial x} - \dfrac{\partial u}{\partial y} \end{matrix} \qquad [3.90]$$

When applied to the components $u, v, w(=0)$ of this flow, we find that the three components of the rotational velocity are at zero.

The flow is not rotational.

5) We directly calculate the two components of the Lagrangian acceleration $\vec{\Gamma} = \dfrac{d\vec{V}}{dt}$ in Cartesian coordinates, for a flat (Ox, Oy) and stagnant flow $(\dfrac{\partial}{\partial t} \equiv 0)$:

$$\Gamma_x = \frac{du}{dt} = u\frac{\partial u}{\partial x} + v\frac{\partial u}{\partial y} \qquad [3.91]$$

$$\Gamma_y = \frac{dv}{dt} = u\frac{\partial v}{\partial x} + v\frac{\partial v}{\partial y} \qquad [3.92]$$

We finally find:

$$\Gamma_x = -A^2 x \qquad [3.93]$$

$$\Gamma_y = -A^2 y \qquad [3.94]$$

6) The result for question (4) could have immediately been predicted. An incompressible flow is derived from a potential and the velocity is a gradient, so its Curl of the velocity is necessarily zero.

The result for question (5) can be directly found by recalling that we have a rotation movement in a block. Each fluid particle therefore has a circular uniform velocity movement $V = \dfrac{1}{r}$. Its acceleration is therefore normal to the trajectory, meaning the radial, which has the standard:

$$\Gamma = \frac{V^2}{r} = \frac{A^2 r^2}{r} = A^2 r \qquad [3.95]$$

We call θ the angle of the radius vector $\vec{r}$ and of the axis Ox. We therefore have:

$$\cos\theta = \frac{x}{r} \qquad [3.96a]$$

$$\sin\theta = \frac{y}{r} \qquad [3.96b]$$

In addition, the two components of $\vec{\Gamma}$ at x and y are written as:

$$\Gamma_x = -\Gamma\cos\theta = A^2 r\frac{x}{r} = -A^2 x \qquad [3.97]$$

$$\Gamma_y = -\Gamma\sin\theta = A^2 r\frac{y}{r} = -A^2 y \qquad [3.98]$$

which is exactly the result found previously.

We note the lesser in the expression of the acceleration components, which express that the acceleration is central, and thus directed toward O.

EXAMPLE 3.5 (Free vortex).–

We imagine that a flat flow, placed in a Cartesian frame, can be described in Eulerian terms by its velocity vector $\vec{V}$ with the components $u\ (x, y)$ and $v\ (x, y)$:

$$u\ (x, y) = \frac{\Gamma}{4\pi}\frac{-y}{\left(x^2 + y^2\right)} \qquad [3.99]$$

$$v\ (x, y) = \frac{\Gamma}{4\pi}\frac{x}{\left(x^2 + y^2\right)} \qquad [3.100]$$

where Γ is a constant whose significance will be revealed later.

1) Show that this flow is possible for an incompressible fluid.

2) Calculate the (Lagrangian) acceleration $\frac{d\vec{V}}{dt}$ of this flow's particle fluids.

3) Determine the trajectories of the fluid particles. How do you name this flow?

Solution:

We can immediately notice that the denominator of the expressions of u *and* v *is equal to the squared distance from the fluid particle to the axis* r. *As a matter of fact,* $r^2 = x^2 + y^2$ *(*r *is the standard of the radius vector* $\vec{r}$, *with components* x, y, z*).*

We can therefore intuit from this point onwards a flow which has a revolutionary symmetry or a relation with a rotation.

1) An incompressible flow has a zero divergence.

Let us see if this is verified.

The flow is clearly stagnant:

$$\frac{\partial u}{\partial x} + \frac{\partial v}{\partial y} = \frac{\Gamma}{4\pi}\left[\frac{\partial}{\partial x}\left(\frac{-y}{x^2+y^2}\right) + \frac{\partial}{\partial y}\left(\frac{x}{x^2+y^2}\right)\right] = \frac{\Gamma}{4\pi}\frac{2yx - 2xy}{\left(x^2+y^2\right)^2} = 0 \quad [3.101]$$

The fluid is definitely incompressible.

2) We directly calculate the acceleration components:

$$\Gamma_x = \frac{du}{dt} = u\frac{\partial u}{\partial x} + v\frac{\partial u}{\partial y} \quad [3.102]$$

$$\Gamma_y = \frac{dv}{dt} = u\frac{\partial v}{\partial x} + v\frac{\partial v}{\partial y} \quad [3.103]$$

$$\Gamma_x = \left(\frac{\Gamma}{4\pi}\right)^2\left[\left(\frac{-y}{x^2+y^2}\right)\frac{\partial}{\partial x}\left(\frac{-y}{x^2+y^2}\right) + \left(\frac{x}{x^2+y^2}\right)\frac{\partial}{\partial y}\left(\frac{-y}{x^2+y^2}\right)\right] \quad [3.104]$$

$$\Gamma_y = \left(\frac{\Gamma}{4\pi}\right)^2\left[\left(\frac{-y}{x^2+y^2}\right)\frac{\partial}{\partial x}\left(\frac{x}{x^2+y^2}\right) + \left(\frac{x}{x^2+y^2}\right)\frac{\partial}{\partial y}\left(\frac{x}{x^2+y^2}\right)\right] \quad [3.105]$$

Once all the calculations have been carried out, it becomes:

$$\Gamma_x = \left(\frac{\Gamma}{4\pi}\right)^2 \frac{-x}{\left(x^2+y^2\right)^3} = \left(\frac{\Gamma}{4\pi}\right)^2 \frac{-x}{r^6} \quad [3.106]$$

$$\Gamma_y = \left(\frac{\Gamma}{4\pi}\right)^2 \frac{-y}{\left(x^2+y^2\right)^3} = \left(\frac{\Gamma}{4\pi}\right)^2 \frac{-y}{r^6} \quad [3.107]$$

At this stage, we notice that the acceleration is carried by the radius vector $\vec{r}$ (position vector of the fluid particle, with components x and y). The direction of this acceleration therefore passes the source of the coordinates O; the sign (–) furthermore shows that the acceleration vector $\vec{\Gamma}$ is directed toward this origin.

3) In order to determine the current lines, we will use the usual relations:

$$\frac{dx}{u} = \frac{dy}{v} \qquad [3.108]$$

or, more practically:

$$v\,dx = u\,dy \qquad [3.109]$$

As the flow is stagnant, these lines will also be the trajectories.

$$\frac{\Gamma}{4\pi}\frac{x}{\left(x^2+y^2\right)}dx = \frac{\Gamma}{4\pi}\frac{-y}{\left(x^2+y^2\right)}dy \qquad [3.110a]$$

which is simplified into:

$$x\,dx = \ -y\,dy \qquad [3.110b]$$

In addition, by way of integration, it gives:

$$\frac{y^2}{2} = -\frac{x^2}{2} + C \qquad [3.111]$$

$$y^2 + x^2 = r^2 = C \qquad [3.112]$$

where C is a constant.

The current lines are therefore the circles centered on the source of the coordinates, with a radius $\sqrt{C}$.

We notice then that the velocity has a module V such as:

$$V^2 = u^2\,(x,y) + v^2(x,y) = \left(\frac{\Gamma}{4\pi}\right)^2 \frac{x^2+y^2}{\left(x^2+y^2\right)^2} = \left(\frac{\Gamma}{4\pi}\right)^2 \frac{1}{r^2} \qquad [3.113]$$

$$V = \frac{\Gamma}{4\pi\,r} \qquad [3.114]$$

The movement of fluid particles is a circular uniform movement, and its acceleration will therefore be normal to the trajectory, meaning radial. We find the results to the question again.

This flow is known by the name "free vortex".

As an exercise, we could calculate its curl of the velocity.

The only component that is not at zero is pointing in the direction of Oz, and therefore, it is perpendicular to the flow's flat line (see [3.32] or [3.90]):

$$rot\,\vec{V}_z = \frac{\partial v}{\partial x} - \frac{\partial u}{\partial y} = \frac{\Gamma}{4\pi}\left[\frac{\partial}{\partial x}\frac{x}{\left(x^2+y^2\right)} - \frac{\partial}{\partial y}\frac{-y}{\left(x^2+y^2\right)}\right] = \frac{\Gamma}{4\pi}\frac{-2x^2-2y^2}{\left(x^2+y^2\right)^2} \quad [3.115]$$

$$rot\,\vec{V}_z = -\frac{\Gamma}{2\pi r^2} \quad [3.116]$$

We note that it is infinite to the source. Nevertheless, the flux of the Curl of the velocity toward a surface containing this source remains finite. We will verify it using Stokes' theorem (see [3.33]):

$$\int_C \vec{V}\,d\vec{l} = \iint_S rot\,\vec{V}.\vec{n}\,dS \quad [3.117]$$

$$\int_C \vec{V}\,d\vec{l} = \frac{\Gamma}{4\pi r}2\pi r = \frac{\Gamma}{2} \quad [3.118]$$

We find that the circulation remains not only finite, but also constant with $\int_C \vec{V}\,dr$. What can be explained by noticing that the Curl of the velocity decreases in $\frac{1}{r^2}$, while the surface of the circle increases in r^2?

NOTE.– This type of flow is the basis of the vortex thread theory. We will not develop this theory further here. We will simply illustrate this remark by specifying that the surface of separation between the two connected flows of perfect uniform fluids with two different velocities can be modeled by an infinity of these cortex threads The instability of the interface can therefore be calculated as the movement of these vortex threads. This opens up to the interfacial instability being processed digitally. The interfacial instability's application is found particularly in atomization problems.

EXAMPLE 3.6 (Non-stagnant compressible flow).–

A flow is described in Eulerian terms by $\vec{V} = (u, 0, 0)$ in Cartesian coordinates.

It verifies:

$$u = \frac{x}{1+t} \tag{3.119}$$

$$\rho = \frac{\rho_0}{1+t} \tag{3.120}$$

where ρ_0 is a constant.

1) Is this flow possible? Is this flow compressible or incompressible, permanent or steady?

2) Calculate the mass m contained at the instant t on the inside of a cylinder in the section S limited by the flat lines $x = 1$ and $x = 3$. Calculate the rate of temporary variation $\frac{dm}{dt}$ of this mass.

3) Using two different averages, calculate the mass flux q_m crossing the cylinder.

4) Calculate the acceleration components of the fluid $\vec{\Gamma} = \frac{d\vec{V}}{dt}$ at each point.

Solution:

NOTE.– *We are faced with an unsteady fluid:* u is a function of t. *Furthermore, the density* ρ *is also explicitly a time function. The fluid is therefore compressible. On the other hand, the flow is mono-dimensionally flat. It is therefore a parallel and non-uniform flow.*

1) In order to be "possible", the flow must at least verify the continuity equation written in non-stagnant terms for a compressible fluid.

$$rot\,\vec{V}_z = \frac{\partial v}{\partial x} - \frac{\partial u}{\partial y} = \frac{\Gamma}{4\pi}\left[\frac{\partial}{\partial x}\frac{x}{\left(x^2+y^2\right)} - \frac{\partial}{\partial y}\frac{-y}{\left(x^2+y^2\right)}\right] = \frac{\Gamma}{4\pi}\frac{-2x^2-2y^2}{\left(x^2+y^2\right)^2} \qquad [3.121]$$

$$\frac{\partial \rho}{\partial t} + \frac{\partial u}{\partial x} = \frac{\partial}{\partial t}\left(\frac{\rho_0}{1+t}\right) + \frac{\partial}{\partial x}\left(\frac{\rho_0}{1+t}\frac{x}{1+t}\right) = \frac{-\rho_0}{(1+t)^2} + \frac{\rho_0}{(1+t)^2} = 0 \qquad [3.122]$$

The continuity equation is therefore verified.

2) The reference volume is:

$$V_{OL} = S(3-1) = 2S \qquad [3.123]$$

The mass that it contains at an instant is therefore:

$$m = \rho S = \frac{2\rho_0 S}{1+t} \qquad [3.124]$$

This mass' variation rate is therefore found by deriving it against the time:

$$\frac{dm}{dt} = \frac{d}{dt}\left(\frac{2\rho_0 S}{1+t}\right) = \frac{-2\rho_0 S}{(1+t)^2} \qquad [3.125]$$

3) The principle of continuity allows us to write that the temporary mass variation results from the difference between the flow rates entering ad emerging from the cylinder in question:

$$\frac{dm}{dt} = \iint_S \vec{V}.\vec{n}\,dS \qquad [3.126]$$

from where two ways of calculating the flow rate's balance sheet emerge:

$$\iint_S \vec{V}.\vec{n}\,dS$$

a) By the value of $\frac{dm}{dt}$ that we already know, we apply the continuity principle:

$$\iint_S \vec{V}.\vec{n}\,dS = \frac{-2\rho_0 S}{(1+t)^2} \qquad [3.127]$$

b) By the direct calculations of the flow rates.

As the velocity is co-linear to the axis Ox, meaning that it is equal to the cylinder's axis. Only the flow rates crossing the faces which are normal to Ox are not zero, so:

$$\iint_S \vec{V}.\vec{n}\, dS = S\left[V(x=1) - V(x=3)\right] = S\frac{\rho_0 1}{1+t} - \frac{\rho_0 3}{1+t} = -\frac{2\rho_0 S}{1+t} \qquad [3.128]$$

which is the value calculated in (a) (equation [3.127]).

4) The Lagrangian acceleration components by the expressions recalled in [3.3] and [3.4], being aware of the fact that the flow is mono-dimensional (v = w = 0), but non-stagnant:

$$\Gamma_x = \frac{du}{dt} = \frac{\partial u}{\partial t} + u\frac{\partial u}{\partial x} \qquad [3.129]$$

$$\Gamma_y = \frac{dv}{dt} = 0 \qquad [3.130]$$

Only the component in x will not be zero:

$$\Gamma_x = \frac{\partial u}{\partial t} + u\frac{\partial u}{\partial x} = \frac{\partial}{\partial t}\frac{x}{1+t} + \frac{x}{1+t}\frac{\partial}{\partial x}\frac{x}{1+t} = \frac{-x}{(1+t)^2} + \frac{x}{1+t}\frac{1}{1+t} \qquad [3.131]$$

$$\Gamma_x = 0 \qquad [3.132]$$

NOTE.– We find that:

$$\frac{du}{dt} = 0 \qquad [3.133]$$

is very different to $\frac{\partial u}{\partial t} = \frac{-x}{(1+t)^2}$ [3.134]

EXAMPLE 3.7.–

In the Eulerian form, a flat flow is given by the components u and v, respectively, following Ox and Oy of its velocity $\vec{V}$:

$$u = \frac{a}{x} \tag{3.135}$$

$$v = -\frac{b}{y} \tag{3.136}$$

1) Is this flow permanent or unsteady?

2) Is it possible to obtain this flow with an incompressible fluid?

3) Give the equation of the current lines of this flow. Would you know to which curve family these current lines belong?

4) Give the components Γ_x and Γ_y of the Lagrangian acceleration $\vec{\Gamma}$ of this flow.

Solution:

1) The expression of velocity components does not contain time. The flow is stagnant (or permanent).

2) If the fluid is incompressible, the flow has a divergence of zero, or:

$$\frac{\partial u}{\partial x} + \frac{\partial v}{\partial y} = 0 \tag{3.137}$$

$$div\vec{V} = \frac{\partial}{\partial x}\frac{a}{x} + \frac{\partial}{\partial y}\frac{-b}{y} = \frac{-a}{x^2} + \frac{b}{y^2} \tag{3.138}$$

This expression is not *a priori* zero in the whole flow. The fluid is therefore compressible.

3) The current lines will be determined based on:

$$v\,dx = u\,dy \tag{3.139}$$

Here:

$$\frac{-b}{y}dx = \frac{a}{x}dy \quad [3.140]$$

which integrates into:

$$\frac{-x\,dx}{b} = \frac{y\,dy}{a} \quad [3.141]$$

$$\frac{y^2}{2a} = \frac{-x^2}{2b} + C \quad [3.142]$$

$$\frac{y^2}{2a} + \frac{x^2}{2b} = C \quad [3.143a]$$

which can be re-written as:

$$\frac{y^2}{2aC} + \frac{x^2}{2bC} = 1 \quad [3.143b]$$

C is a constant. The curves have a conically shaped equation. This is a network of ellipses, of axes Ox and Oy . The lengths of the half-axes according to these two coordinate axes are, respectively, $2aC$ and $2bC$.

4) For a permanent flow, the bi-dimensional flat line, the Lagrangian acceleration r $\vec{\Gamma}$, is written (following [3.3] and [3.4]) as:

$$\Gamma_x = \frac{du}{dt} = u\frac{\partial u}{\partial x} + v\frac{\partial u}{\partial y} \quad [3.144]$$

$$\Gamma_y = \frac{dv}{dt} = u\frac{\partial v}{\partial x} + v\frac{\partial v}{\partial y} \quad [3.145]$$

or

$$\Gamma_x = \frac{du}{dt} = \frac{a}{x}\frac{\partial}{\partial x}\left(\frac{a}{x}\right) - \frac{b}{y}\frac{\partial}{\partial y}\left(\frac{a}{x}\right) = \frac{-a^2}{x^3} \quad [3.146]$$

$$\Gamma_y = \frac{dv}{dt} = \frac{a}{x}\frac{\partial}{\partial x}\left(\frac{-b}{y}\right) - \frac{b}{y}\frac{\partial}{\partial y}\left(\frac{-b}{y}\right) = \frac{b^2}{y^3} \quad [3.147]$$

We notice that the velocity and the acceleration become infinite on the axes.

This somewhat "academic" exercise refers to a "hypothetical" flow, which cannot be materially defined everywhere at each state of cause. This situation is also produced in flows issued from a punctual source where the flow rate's conservation implies an infinite velocity on the point's "zero" section.

EXAMPLE 3.8.–

The bi-dimensional velocity field is given by:

$$u = x^2 y + y^2 \quad [3.148]$$

$$v = x^2 - xy^2 \quad [3.149]$$

1) Show that this flow is stagnant and incompressible.

2) Determine its current function and its Curl of the velocity. What do you notice?

Solution:

1) The expressions of u and v do not contain time. Therefore, the flow is permanent, or steady.

Furthermore, we notice that the flow is flat and bi-dimensional.

In order to be incompressible, the velocity components must verify:

$$div \vec{V} = 0 \quad [3.150]$$

For a flat and bi-dimensional flow:

$$\frac{\partial u}{\partial x} + \frac{\partial v}{\partial y} = 0 \quad [3.151]$$

For this flow:

$$\frac{\partial}{\partial x}(x^2 y + y^2) + \frac{\partial}{\partial y}(x^2 - xy^2) = 2xy - 2yx = 0 \quad [3.152]$$

Thus, we verify that the continuity is fulfilled by this incompressible flow.

2) The stream function $\Psi(x,y)$ is defined by:

$$u = \frac{\partial \Psi}{\partial y}; \quad v = -\frac{\partial \Psi}{\partial x} \qquad [3.153]$$

or:

$$\frac{\partial \Psi}{\partial y} = x^2 y + y^2; \quad -\frac{\partial \Psi}{\partial x} = x^2 - xy^2 \qquad [3.154]$$

which is resolved by successively integrating each equality:

$$\Psi(x,y) = \frac{x^2 y^2}{2} + \frac{y^3}{3} + f(x) \qquad [3.155]$$

$$\Psi(x,y) = \frac{-x^3}{3} + \frac{x^2 y^2}{2} + g(y) \qquad [3.156]$$

$f(x)$ and $g(y)$ are two unknown functions. The comparison of the two previous lines leads to:

$$f(x) = \frac{-x^3}{3} \qquad [3.157]$$

$$g(y) = \frac{y^3}{3} \qquad [3.158]$$

From this, the expression of $\Psi(x,y)$ results, which is defined as a constant near:

$$\Psi(x,y) = \frac{x^2 y^2}{2} + \frac{y^3 - x^3}{3} + C \qquad [3.159]$$

This expression in principle enables us to find the equation of the current lines for them. We will not perform a complex calculation here, as it is not required.

$$\Psi(x,y) = \frac{x^2 y^2}{2} + \frac{y^3 - x^3}{3} = Cte \qquad [3.160]$$

The velocity field depends only on x and y, and has no component w following Oz.

The curl of the velocity $rot\vec{V}$ of the velocity field may possibly have a component in z:

$$rot\,\vec{V}_z = \frac{\partial}{\partial x}\left(x^2 y + y^2\right) - \frac{\partial}{\partial y}\left(x^2 - xy^2\right) = 2xy - 2yx = 0 \qquad [3.161]$$

which was foreseeable. This flow is incompressible and it had a velocity potential. A flow with a velocity potential is not rotational.

4

Dynamics of Inviscid Fluids

4.1. Introduction

The Bernoulli theorem can be used quickly and easily, and is essential in fluid mechanics. Some would even say that, along with hydrostatics, it summarizes the field. As it were, a general approach to non-viscous fluids involves the resolution of equations of fluid mechanics, where the terms of viscosity have been removed.

In this case, the fluid is said to be "perfect". There is no such thing as an intrinsically perfect fluid, and it is instead the equations that become "perfect", for the convenience of the physicist. These equations contain information that theoretically allows for the complete determination of the flow, notably the flow kinematics. Here we focus on a more limited domain, with an in-depth study of the use of the Bernoulli theorem, while remaining fully aware that this does not cover the full topic!

4.2. The Bernoulli theorem: proof

While the Bernoulli theorem is highly applicable, it also has considerable limitations, if only by the hypotheses that underlie it.

To properly understand the reach of this theorem, we have thought it pertinent to provide a proof, less classical perhaps than most, but which clearly shows its relation to the fundamental principle of dynamics. Moreover, another relation can be made apparent, and is too often neglected by students (and sometimes textbooks).

One of the most common proofs seen in textbooks is the application of the theorem of kinetic energy, and of course the direct logical entailment of general equations without the terms representing viscosity. The latter proof is the most elegant one, which is based on a treatment of the general equations. For a better physical understanding of the application of dynamics by all those reading, we prefer to use the following version.

Let us first comment on the spelling of the name, although this is secondary to the physics. The "standard" spelling used in this work is "Bernoulli". In Francophone countries, the "lli" at the end is pronounced like a yii, despite the absence of an "i" as in "illi". This is in fact the Spanish pronunciation of the name, which is paradoxical as the Bernoulli family were Dutch refugees living in Switzerland. In some textbooks, "Bernoully" or "Bernouilly" are still seen, presumably as a result of the authors' quest for historical authenticity. A rather random quest, as in the 18th Century, the same name could be spelled in three different ways on the same page, even in a royal decree.

The hypotheses required for the application of the Bernoulli theorem in its classical form are:

a) the fluid must be *incompressible*, (synonym: density ρ is constant);

b) the fluid is *perfect* (no tangential forces);

c) the flow is *permanent* (resulting in a line of the current being also a trajectory);

d) the *volume forces are limited to gravity*.

We shall later go beyond this overly restrictive hypothesis, by extending the theorem to all of the derivative forces of a potential. O is a point along a line of a current/trajectory. We shall write the fundamental principle of the dynamics of a fluid particle passing through this point at a time t. We shall also project this fundamental principle of dynamics onto two axes of a very particular reference frame. This reference frame $Oxyz$ has origin O, around which three axes are built as follows:

– axis Ox is tangential to the trajectory at O;

– axis Oy is normal to the line of current;

– the third axis Oz creates the direct orthonormal system with Ox and Oy.

Such a system of axes is only used between the moment $t = 0$, where the particle passes through O and a later moment dt. Such a "single-use" system of axes is called an intrinsic axis system. The fluid particle is represented by an infinitely small parallelepiped with sides dy, dx and dz that are carried by the three axes. Its volume and mass are therefore $dx\,dy\,dz$ and $\rho dxdydz$, respectively.

Velocity $\vec{V}$ of the particle has a single non-nil component, along Ox, which is a modulus of this velocity, denoted by V. The (Lagrangian) acceleration $\frac{d\vec{V}}{dt}$ of this particle at time t has two components:

Along Ox:

$$\left.\frac{dV}{dt}\right|_x = \frac{dV}{dx}\;\frac{dx}{dt} = V\frac{dV}{dx}. \qquad [4.1]$$

Through this small calculation, a derivative with respect to time is transformed into a spatial derivative. We note that this in fact represents a shift from a Eulerian form of writing to a Lagrangian form.

Along Oy:

$$\left.\frac{dV}{dt}\right|_y = \frac{V^2}{R} \qquad [4.2]$$

where R is the curvature radius of the trajectory.

Here we use a result from the kinematics of a material point ("central" acceleration of a planetary movement, for example) without proof. In the case where the trajectory is a circle, R is a constant.

Along Oz, the acceleration component is equal to zero.

Let us call θ the angle made at O by the tangent to the line of the current/trajectory with the horizontal. The forces applied to the elementary parallelepiped particle with sides dx, dx and dz are:

– the surface forces, which, assuming the fluid is perfect, are limited to the forces of pressure, normal to the six faces of the particle;

– the volume forces, limited to gravity.

Let us add an additional axis OZ, which is vertical with a bottom-up direction and does not belong to the axis system $Oxyz$. *This unusual procedure is only used to show the altitude of the fluid particle.* The position of the origin of this axis is unchanged, since, as seen later, only the changes of Z are relevant. The fundamental principle of dynamics is written as:

$$\rho dxdydz\frac{d\vec{V}}{dt}=d\vec{F}_{Volume}+d\vec{F}_{Surface} \qquad [4.3]$$

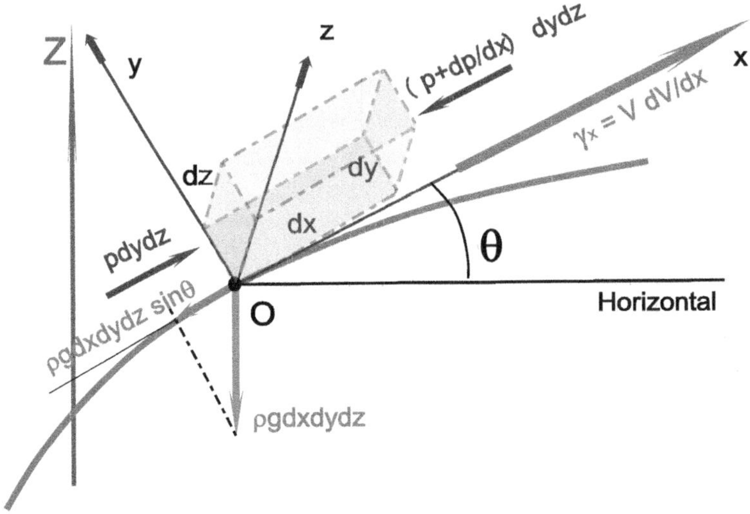

Figure 4.1. *Fundamental principle of dynamics projected onto the Ox axis with intrinsic coordinates*

We now project it onto the Ox axis at time $t=0$ when the fluid particle passes through it:

$$\rho\, dxdydz\; V\frac{dV}{dt}=dxdz\left[p-\left(p+\frac{\partial p}{\partial y}dy\right)\right]-\rho\, dxdydz\, g\sin\theta \qquad [4.4]$$

In this projection, we have taken into account the angle θ of the tangent to the line of current with the horizontal.

Moreover, only the forces of pressure on $dy\,dz$ have a non-nil projection on Ox.

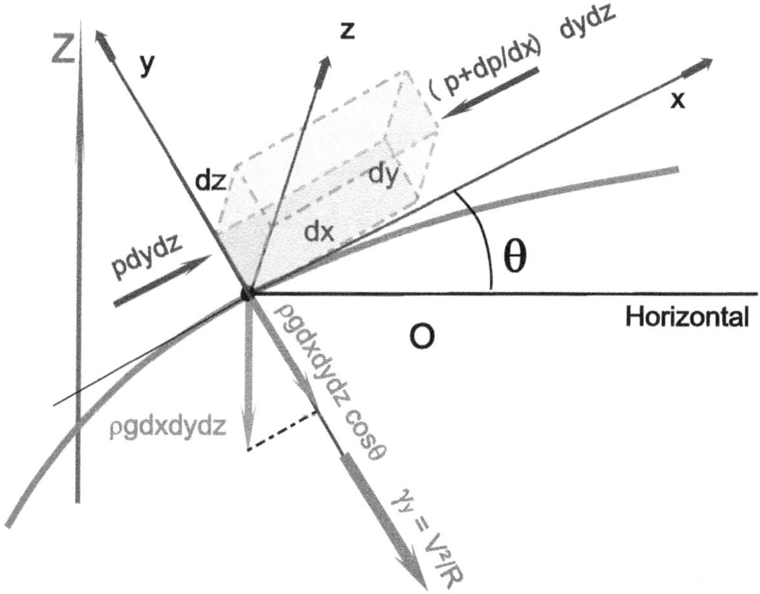

Figure 4.2. *Fundamental principle of dynamics projected onto the Oy axis with intrinsic coordinates*

We now project the fundamental principle of dynamics onto the Oy axis. Using an analogous method, we obtain:

$$\rho\, dxdydz \frac{V^2}{R} = dxdz\left[p - \left(p + \frac{\partial p}{\partial y}dy\right)\right] - \rho\, dxdydz\, g\cos\theta \qquad [4.5]$$

Moreover, we note that an infinitely small movement of a point along the line of the current results in two variations of x and y, dx and dy, which are linked to the corresponding variation in altitude, dZ through:

$$\frac{\partial Z}{\partial x} = \sin\theta \qquad [4.6]$$

$$\frac{\partial Z}{\partial y} = \cos\theta \qquad [4.7]$$

Let us focus here on the important distinction between dz, tied to the intrinsic axis system, which is not considered here, and dZ, which represents the variation in altitude.

Projection onto Ox, after some manipulating, results in:

$$dxdydz\,\rho V\frac{dV}{dx} = dxdydz\left(-\frac{\partial p}{\partial x}dx - \rho g\sin\theta\right) = dxdydz\left(-\frac{\partial p}{\partial x}dx - \rho g\frac{\partial Z}{\partial x}\right) \quad [4.8]$$

$$\rho V\frac{dV}{dx} + \frac{\partial p}{\partial x}dx + \rho g\frac{\partial Z}{\partial x} = 0 \quad [4.9]$$

which, by gathering the derivatives in x, results in:

$$\frac{\partial}{\partial x}\left(\rho\frac{V^2}{2} + p + \rho gZ\right) = 0 \quad [4.10]$$

The Bernoulli theorem can be stated as:

Along a streamline (or in other words from one point of this line to another), the quantity $\rho\frac{V^2}{2} + p + \rho gZ$ is constant [4.11]

Projection onto Oy, after multiple analogous manipulations, also results in the following:

$$\rho\, dxdydz\frac{V^2}{R} = dxdz\left[p - \left(p + \frac{\partial p}{\partial y}dy\right)\right] - \rho\, dxdydz\, g\cos\theta \quad [4.12]$$

$$\rho\, dxdydz\ \frac{V^2}{R} = dxdydz\left(\frac{\partial p}{\partial y} + \rho g\frac{\partial Z}{\partial y}\right) \quad [4.13]$$

$$\rho\frac{V^2}{R} = \frac{\partial}{\partial y}(p + \rho gZ) \quad [4.14]$$

We can see that the curvature radius R of the current line appears at the point chosen as the origin. When the geometry of this line is circular, R is the radius of the circle and a classical formula can be used. In the case of more complex curvature geometry, or even a three-dimensional one, the acceleration component, located in the osculator plane, is written as:

$$\gamma_y = \frac{V^2}{R} \quad [4.15]$$

One case in particular leads to an important result. In the areas where the streamlines are parallel lines, R is infinite, resulting in:

$$\frac{\partial}{\partial y}(p+\rho gZ)=0 \quad [4.16]$$

In other words, on any plane perpendicular to the streamlines of a parallel flow, the quantity:

$$p_G = p + \rho gZ \text{ is constant.} \quad [4.17]$$

The quantity p_G, often called the *generating pressure*, is the quantity that remains constant throughout the whole of the immobile fluid. Here again it can be noted that Newton's first law, or the inertia principle, can be viewed as a consequence of the second law, which is the fundamental principle of dynamics. We note that the flow needs only to be parallel and that it does not need to be uniform. It is sometimes said that on a plane that is perpendicular to the lines of current of a parallel flow, the distribution of the pressures is hydrostatic. This is best used with caution, as the plane in question is not usually vertical. Here we are dealing with fluid dynamics, and we will often use this important property, which moreover is regularly applicable to real fluids. By going back to the proof, we can now see that the forces of the tangential surface need to have no component along the Oy axis, and this is often verified to be the case.

4.2.1. *What to retain*

To resolve the following problems, it is important to remember that it is the projection of the fundamental law of dynamics over two axes that provide two important results, stemming from the four preliminary hypotheses used in the proof:

a) Along the same line of the current, the quantity $p_T = \rho\frac{V^2}{2} + p + \rho gZ$ or the total pressure is constant.

b) On any plane normal to a parallel flow zone, the quantity $p_G = p + \rho gZ$, or generating pressure, is constant. This property remains true for a real fluid when the viscosity forces are tangential to the lines of current (which is often verified).

Beyond this traditional statement, by going back to the previous proof, the reader can attempt to demonstrate (and also test their understanding) that in the case where the volume forces are no longer limited to gravity but still are derived from the potential $\phi(x,y,z)$, the constant quantity along the line of the current becomes $\rho\frac{V^2}{2}+p+\rho\,\phi$. This very general formula also includes the case of gravity for which $\phi=gZ$. The potential $\phi(x,y,z)$, as always for the volume forces, is defined by unit of mass.

This calls for an "extended" Bernoulli theorem:

a) Along the same line of current, the total pressure $p_T=\rho\frac{V^2}{2}+p+\rho\phi$ is constant.

b) On any plane normal to a parallel flow zone, the generating pressure $p_G=p+\rho\phi$ is constant. This property remains true for a real fluid when the viscosity forces are tangential to the lines of the current (which is often verified).

4.2.2. *Energetic interpretation of the Bernoulli theorem*

This interpretation provides a useful aid in understanding the *physical meaning of this theorem.*

The quantity preserved along the line of the current, referred to as the "total pressure", $p_T\,\rho\frac{V^2}{2}+p+\rho\phi$ has the dimension of energy per unit of volume. It is the sum of three volumic energies:

a) A kinetic energy per unit of volume: $\frac{\rho V^2}{2}$.

b) A potential energy per unit of volume: ρgz (or $\rho g\phi$ if the volume forces have more than one component).

c) Energy stored as pressure per unit of volume: $\frac{\rho V^2}{2}$.

This concept can be better understood by imagining a gas in a cylindrical piston system. By compressing this gas, the compression work W provided is stored as internal energy.

The total pressure P_T can therefore be written as the sum of a *"volumic" kinetic energy* and a "generating pressure" p_G *that is the sum of the two "volumic" potential energies* p *and* $\rho g\phi$:

$$p_T = \rho\frac{V^2}{2} + p + \rho\phi = \rho\frac{V^2}{2} + p_G;\ p_G = p + \rho\phi \qquad [4.18]$$

4.2.3. *Physical interpretation of the Bernoulli theorem*

This interpretation is simple: when the perfect fluid is "thrown", no forces that are parallel to the movement of its fluid particles can be applied to it. As a result, no surface forces can work. The sum of the kinetic and potential energies of each fluid particle is then invariable. It is important to note, however, that the volume forces are working. However, this work is included in the variation of a potential ϕ from one point of the flow to another. A good analogy for this is a high-speed train racing along at 300 km.h^{-1}. If the friction on the rails is considered to be negligible, no energy is expended during a turn along the line. The reactive forces of the rail are normal to the movement and therefore do not work.

4.2.4. *"Constant energy" flows*

In many given problems, the flow comes from a reservoir, which is a space within which its velocity is nil. In this sense, it is necessary to conceive streamlines that link a point of the flow where the velocity is not nil to a point of the reservoir where the velocity vector is meant to exist, but which has a norm that is equal to zero.

The most important is that the fluid in the reservoir verifies the fundamental theorem of statics (here hydrostatics) and therefore that $p + \rho gz$ or $p + \rho\phi$ is constant. V being equal to zero, $\dfrac{\rho V^2}{2} + p + \rho gz \equiv p + \rho gz$ is constant inside the reservoir, and remains such throughout the entire flow. This type of flow is called a *constant energy* flow. It is also seen in the case where at least one zone of the flow is uniform. Indeed, in a parallel flow, any plane that is perpendicular to the lines of the current has $p + \rho gz$ or $p + \rho\phi$ that is constant. Moreover, assuming the flow has a constant velocity, the zone of uniform flow is such that $\frac{\rho V^2}{2} + p + \rho gz$ or $\frac{\rho V^2}{2} + p + \rho\phi$ is constant throughout. When the lines of the current eventually separate, this quantity retains the same value on every line, and therefore throughout the entire flow.

4.3. Applications of the Bernoulli theorem

4.3.1. *Methodology for the resolution of a problem using the Bernoulli theorem*

While very practical and easy to use, the Bernoulli theorem also has some limitations. The Bernoulli theorem can be applied to a streamline, but can in no case be used to determine it. As a result, the kinematic elements of the flow must be known, or at least "imagined".

In some scenarios, these kinematic elements are provided. In the majority of cases, the physicist must use his or her common sense. A frequent case is that of a reference point attached to a mobile object (aircraft, land vehicle, etc.). In this case, the "far-off" liquid ahead of the mobile element undergoes a uniform flow in relation to the reference point. Let us remember this notion of "far-off". As a general rule, flows that are close to a mobile object are the result of the interaction between this "far-off" uniform flow and the mobile object acting as an obstacle. This results in a flow with curved current lines, leading to a complex spatial distribution of velocities and especially of pressures. With very little information available on these "pathological" zones, it is best to avoid choosing points within them when writing the Bernoulli theorem.

As a general rule, the application of the Bernoulli theorem relies implicitly on the following statement:

– At every point of a flow, three parameters are key: the applicate Z (which is the altitude, on a vertical axis going bottom-up, thus reducing the problem to one spatial dimension), the static pressure P and the velocity modulus V. The first two parameters can often be condensed into P_G.

The OZ axis is vertical when the volume forces are limited to gravity. When additional forces deriving from a potential are involved, another favored axis can be defined. Then, Z is no longer an altitude.

There is always at least one flow point, A, where all three parameters are known, and B, which is often the subject of the problem, where only two parameters are known. The solution then involves writing the Bernoulli theorem between A and B. The main advantage of the Bernoulli theorem is that its only constraint requires A and B to be on the same current line. No information is required regarding the geometry of these current lines (not more than on the distribution of pressures and velocity in these zones) between points A and B. In practice, all that is needed is to have enough information on the zones containing A and B, and to be sure that there is indeed a current line linking A and B, which is usually not too problematic. Depending on the problem studied, the parameter determined by application of the theorem is either a velocity, a pressure or even an applicate.

The Bernoulli theorem cannot be applied between two points of two different fluids. In any case, incompressibility would not be verified, even with liquids of generally different densities. This is also obviously not possible if A and B are located in two parts of a same fluid separated by an impermeable wall!

The problems provided here can be separated into two categories:

a) Elementary problems.

b) Draining and filling.

A common problem involves the filling or draining of a reservoir. We will come across various types. In some problems, the filling is carried out at a constant fluid velocity. A distinction can be made depending on whether the system in which the operation is being carried out is fixed in relation to the ground, or whether it is mobile. In other problems, the fluid velocity during filling is variable over time. Finally, we shall provide several "synthetic" problems, which call on several of these categories.

We shall first cover simple examples, where the main objective is to determine either a velocity, a cote or a pressure, or several parameters at a time.

Determining velocity: In several problems, the goal is to determine the velocity.

EXAMPLE 4.1 (Angel Falls).–

Throughout the entire problem, water is considered a perfect fluid. Angel Falls is located in Venezuela and is one of the highest waterfalls in the world. Its waters fall freely from a height $h = 807\ m$.

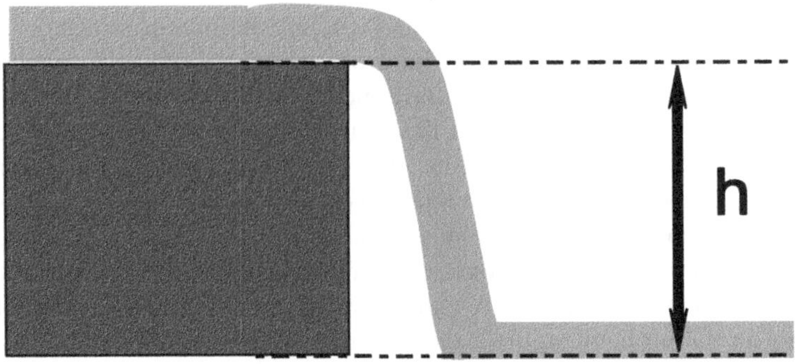

Figure 4.3. *Diagram of a waterfall*

1) By assuming that the flow in the waterfall is uniform and continuous (meaning that the falling water body does not fragment into different filaments, which is perhaps not very realistic), determine the velocity of the water as it reaches the bottom of the fall. Assume the flow velocity at the top of the fall to be negligible.

2) Use the same process to determine the velocity of the flow in Niagara Falls, which is 59 m high, and compare with the previous value.

Contrary to popular belief, Niagara Falls is not the highest in the world. They are, however, among those with the highest flow rate (6,962 $m^3.s^{-1}$).

Solution:

1) Here the Bernoulli theorem must be used, and the points chosen must be: point A at the free surface of the flow upstream of the falls, and point B at the foot of the falls.

The sides are taken on a vertical axis with an upward direction. The origin is at the base of the fall (ground level):

$$\rho\frac{V_A^2}{2} + p_A + \rho g z_A = \rho\frac{V_B^2}{2} + p_B + \rho g z_A \qquad [4.19]$$

There is:

$$p_A = p_a;\ p_B = p_a;\ z_A = H;\ z_B = 0;\ V_A = 0 \qquad [4.20]$$

As a result, the velocity at the base of the fall is:

$$V_B = \sqrt{2gh} \quad [4.21]$$

This formula is the same as the one used in the case of a free fall of a material point from a height h. This form is known as the Torricelli formula, which is not surprising. On a historical note, as early as 1698, Newton studied in "Principia" the case of a falling jet of liquid using the laws of falling bodies.

For Angel Falls, the velocity is determined as:

$$V_B = 125,8\, m.s^{-1} \quad [4.22]$$

2) For Niagara Falls, the velocity at the base of the falls is:

$$V_B = 34\, m.s^{-1} \quad [4.23]$$

EXAMPLE 4.2 (Die Hard with a Vengeance).–

In "*Die Hard with a Vengeance*" (Die Hard 3), Bruce Willis finds himself in a CA truck (see Figure 4.4). This truck drives into an underground pipeline T that is normally separated from a dam reservoir L by a bulkhead (not represented in the diagram). The terrorists have blown up the bulkhead to get in the way of the protagonist. Consequently, the reservoir starts to drain into the pipeline, and at the precise moment we are looking at, Bruce Willis narrowly escapes the resulting flow in his truck. The movie lets us see the speedometer, which is showing 60, presumably miles per hour.

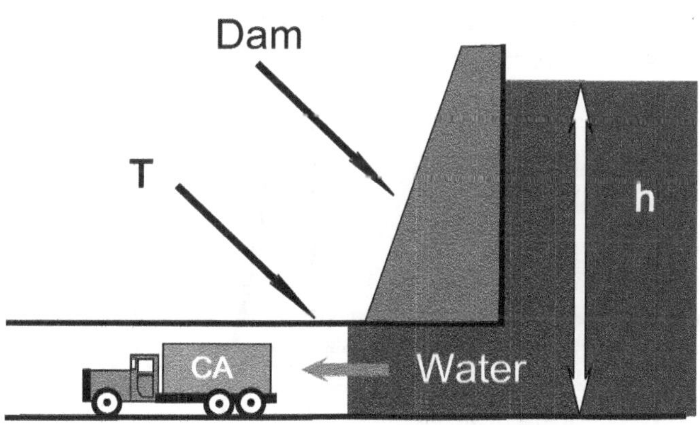

Figure 4.4. *Die Hard 3: a tricky situation*

The question here is to determine whether this speed of 60 miles an hour for the water pursuing 1 h is a credible figure. To answer this, the depth of the water in the reservoir threatening Bruce Willis and his truck can be determined. The water is assumed a perfect fluid. The flow of the water within the tube is also assumed uniform. *It is worth noting that a mile is equal to 1609 m.*

Solution:

The problem is similar to that of the Angel Falls. In this case, we know the water velocity at the base and instead we are looking for the length to the high point of the reservoir:

$$V_B = \frac{60*1609}{3600} = 26{,}82\ m.s^{-1} \quad [4.24]$$

$$V_B = \sqrt{2gh} \quad [4.25]$$

$$h = \frac{V_B^2}{2g} = 36{,}65\,m \quad [4.26]$$

This height is entirely plausible. By considering uniform flow, this implies that the altitude varies little in T considering h. Given the average size of a truck, this hypothesis is rather tenuous.

EXAMPLE 4.3 (Parameters of a hydraulic dam).–

A dam is used to create a reservoir with a depth (H) of 50 m. A pipeline C, with a diameter (D) of 50 cm, whose axis is located at h = 40 m below the level of the free surface of the reservoir crosses the dam to power a hydroelectric turbine. Atmospheric pressure is $p_a = 1$ bar.

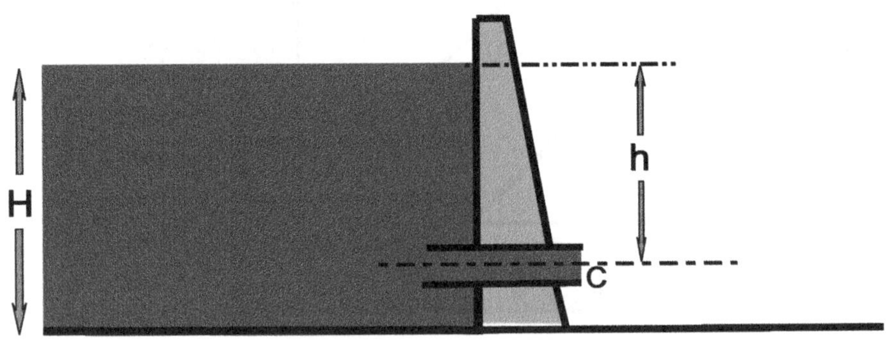

Figure 4.5. *Hydraulic dam*

1) What is the pressure p_f at the bottom of the reservoir?

2) Assuming the water in the reservoir is a perfect fluid, what is the flow rate q_v of the water that powers the turbine?

3) What energy output, expressed in kW, is available to power this turbine?

Solution:

1) This is a question related to hydrostatics

$$p_f = p_a + \rho g H \quad [4.27]$$

$$p_f = 10^5 + 1000 * 9,81 * 50 = 5,905.10^5\ Pa = 5,9\,bar \quad [4.28]$$

2) By applying the Bernoulli theorem between a point S of the surface of the reservoir and a point T on the same current line at the pipe exit:

$$p_a + \rho g z_S + \rho \frac{V_S^2}{2} = p_a + \rho g z_T + \rho \frac{V_T^2}{2} \quad [4.29]$$

with:

$$V_S = 0;\ z_S - z_T = h;\ V_T = \sqrt{2gh} \quad [4.30]$$

V_T is the outflow velocity at the exit of the pipe. The flow volume can be determined easily:

$$V_T = 28\,ms^{-1} \quad [4.31]$$

$$q_V = \pi \frac{D^2}{4} V_T = 5,5\,m^3\,s^{-1} \quad [4.32]$$

3) The energy output is the amount of energy exiting the pipe per unit of time. This is also the hydraulic power provided to any possible electrical generator (turbine + dynamo or alternator) located downstream. For 1 kg of fluid, the kinetic energy is $e_C = \frac{V_T^2}{2}$. The mass flow is $q_m = \rho V_T$. The energy output or hydraulic power is:

$$P_H = q_m\, e_C \quad [4.33]$$

$$P_H = 1,000 * 5,5 * \frac{(28)^2}{2} = 2,156\quad MW \quad [4.34]$$

EXAMPLE 4.4 (Water jet cutting supply).–

A simple water jet cutting system is composed of a reservoir filled with water at a pressure p_R that must be determined. A nozzle with a diameter $d = 5\,mm$ is linked to this reservoir.

1) The desired jet velocity is $U_j = 900\,m.s^{-1}$, when the nozzle is more or less at the level of the reservoir. What must be the value of p_R, expressed in bars? The density of water is $\rho = 1{,}000\ kg.m^{-3}$. Atmospheric pressure is $p_a = 1\,bar$.

2) A diver is working at sea at a depth of $h = 20\,m$, while the reservoir remains on land. Show that the cutting time is not significantly altered.

Solution:

1) In this example, the velocity is mainly determined by the pressure of the reservoir.

By writing out the Bernoulli theorem between a point A of the reservoir and a point B of the jet, both located on the same line of current, we obtain:

$$p_A + \rho g z_A + \rho \frac{V_A^2}{2} = p_B + \rho g z_B + \rho \frac{V_B^2}{2} \tag{4.35}$$

Clearly:

$$p_A = p_R;\ \ z_A = z_B;\ \ V_A = 0;\ \ V_B = V_j \tag{4.36}$$

Resulting in:

$$p_R = \rho \frac{V_j^2}{2} + p_a \tag{4.37}$$

$$p_R = 1{,}000 * \frac{900^2}{2} + 10^5 = 4{,}051.10^8\ \ Pa = 4{,}051\ \ bar \tag{4.38}$$

2) As an order of magnitude, it can be noted that a depth of 20 m corresponds to an increase in pressure of $\Delta p = \rho g h = 1{,}000 * 9{,}81 * 20 = 1{,}96.10^5\,Pa$, which is

very small compared with p_R. Therefore, the velocity of the jet will not be significantly impacted. It can be calculated explicitly:

$$p_R + 0 + = \rho \frac{V_j^2}{2} + p_a + \rho g h \qquad [4.39]$$

$$V_j = \sqrt{2.\frac{p_R - (p_a + \rho g h)}{\rho}} = \sqrt{2\frac{4{,}501.10^8 - 2{,}96.10^5}{1{,}000}} = 899{,}8 \ \ m.s^{-1} \qquad [4.40]$$

Resulting in $V_j = 899{,}8 m.s^{-1}$ which is equal to $V_j = 899{,}8 m.s^{-1}$ to the nearest 2.10^{-4}.

EXAMPLE 4.5 (Measurement by Pitot tube).–

Since the early days of aviation, the airspeed of airplanes has been measured using a Pitot tube. This device is essentially a tube with a rounded end. At the tip of this tube, there is a channel that measures the pressure p_S that exists inside this rounded end.

Over time, various methods for measuring this pressure have been used, from U-tubes to modern probes. Here we look at the value of the pressure measured on its own. The tube is placed on the wing of the plane, which moves in relation to the surrounding air at a velocity of V_{rel}. p_a is the atmospheric pressure ahead of the plane. An airplane moves through the air at an atmospheric pressure of $p_a = 0{,}8 bar$. The airplane is working against wind with a speed of $V_w = 150 km.hr^{-1}$ in relation to Earth. The Pitot tube measures a pressure of $p_S = 0{,}859 bar$. The density of air is $\rho = 0{,}96 kg.m^{-3}$ at the altitude in question. What is the speed of the airplane V_{plan} in relation to Earth?

Solution:

The Pitot tube is taken as a fixed referential. The tube is therefore placed in a uniform airflow with a velocity of V_{rel}. Given the orders of magnitude considered and difference of probably velocities, we can assume the air to be incompressible.

To know the atmospheric pressure p_a, Prandtl decided to equip the tube with a double cover with a lateral orifice located quite far back from the extremity S of the tube. At that location, the flow has returned to being parallel, and the velocity has practically returned to V_{rel}. The static pressure is therefore p_a at the orifice in question.

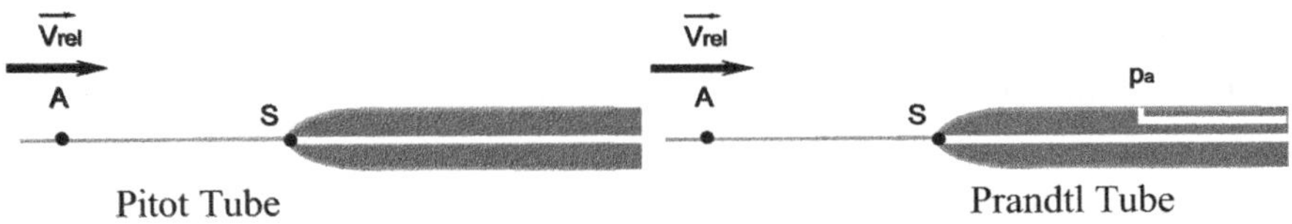

Figure 4.6. *Pitot tube and Prandtl tube*

As the air is assumed to be a perfect fluid, a kinematic hypothesis is taken that can sometimes be troubling for the reader. Along the current line that reaches the point S, which is the orifice of the tube, the assumption is made that the fluid stops on the tube, and then regains its speed later.

We could therefore write out the Bernoulli theorem between a point A located far upstream of the flow, where the velocity is V_{rel}, and the point S, where the velocity is equal to zero:

$$\rho\frac{V_{rel}^2}{2} + p_a + \rho g z_A = \rho\frac{V_S^2}{2} + p_a + \rho g z_S = \rho\frac{V_S^2}{2} + p_S + \rho g z_S \qquad [4.41]$$

The change in the terms of gravity for this light gas is negligible from one point to another. Therefore:

$$\rho\frac{V_{rel}^2}{2} + p_a = p_S \qquad [4.42]$$

$$V_{rel} = \frac{2(p_S - p_a)}{\rho} \qquad [4.43]$$

The relative velocity in relation to the air is:

$$V_{rel} = \frac{2(8.59.10^4 - 8.10^4)}{0,96} = 110,9\,m.s^{-1} = 399\,km.hr^{-1} \qquad [4.44]$$

Assuming the airplane is moving in the same direction as the wind at 150 km.h^{-1}, this velocity must be added in order to reach the groundspeed:

$$V_{av} = 399 + 150 = 449\,km.hr^{-1} \qquad [4.45]$$

NOTE.– The influence of the wind speed on the groundspeed is far from negligible, even for jetliners. During intercontinental trips, differences of more than an hour are commonly seen depending on the direction of travel. The groundspeed is more easily obtained using geolocalization. However, the relative velocity of the airplane in relation to the surrounding air is important for the flight data, if only in terms of keeping the aircraft aloft. Recent disasters have shown the terrible effects caused by a malfunction of a Pitot tube (when ice gets into the tube, for example) if it is not detected in time.

4.3.2. *Determining an applicate*

In other, also quite basic problems, the objective is to determine the length of an applicate (or more concretely, an altitude). This is the case of the jet problems.

EXAMPLE 4.6 (Height of a jet).–

A liquid jet is a flow that is not limited by solid surfaces. A jet can be created from the device represented below. The piston p, with a diameter of D, is quite heavy and has a mass of m. It is suitably watertight, as required. This piston descends *very slowly* into the cylinder C, which is always filled with liquid, excluding any gas. This liquid is a perfect fluid with density ρ. T is an orifice with a small diameter d, which is very small compared with D. It exits just above the piston P.

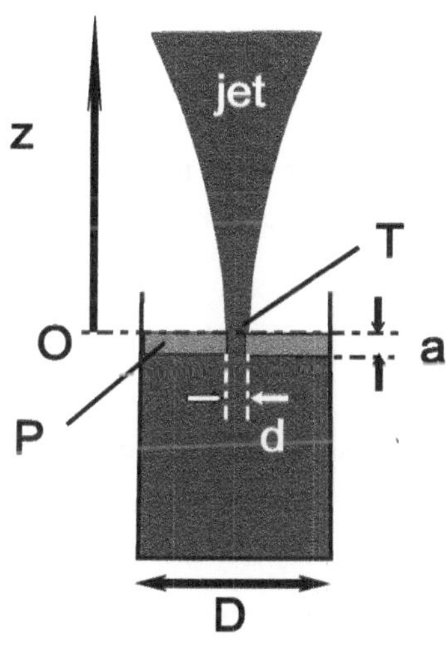

Figure 4.7. *Creating a jet*

1) A jet is created at the exit T.

1.1) Show that the velocity of this jet decreases with altitude. Show that theoretically, at the upper extremity of the jet, its cross-sectional area must be infinite. We shall not worry about this here.

1.2) By applying the Bernoulli theorem between the lower face of the piston and a point of jet with an applicate z, find the relationship between the velocity V_Z of the liquid and the applicate z of a point of the jet. Give the expression of the cross-sectional area of this jet versus z. The origin of the different " z " s is taken at the level of the top of the piston. The thickness of the piston is a. *Do not forget that atmospheric pressure also applies above the piston and that its effects are therefore be added to those caused by the weights of the piston.*

2) We can predict that the jet will escape upward. At a certain altitude, it can be assumed that the liquid will fall back down without affecting the upstream flow of the jet.

What is the value of the height z_0, expressed in relation to the piston, that the jet will reach?

NOTE.– This is a simple model. In reality, the jet does not maintain its integrity; it breaks into filaments and droplets. Moreover, air resistance plays a considerable role.

Solution:

1.1) The principle of continuity ensures that the flow in the jet will maintain a constant value. If $S(z)$ is the cross-sectional area of the jet at a given applicate (altitude), we get:

$$q_m = \rho S(z) V(z) \quad [4.46]$$

Along the current line in the ascending jet, the following expression applies, and is constant for every point of this line:

$$p_T = \rho \frac{V^2(z)}{2} + p_a + \rho g z \quad [4.47]$$

The key to jet problems lies in the fact the flow is "free". Its borders are always in contact with the atmosphere (unlike what happens in a pipeline). For reasons of continuity, the pressure inside the jet is therefore independent of the applicate and is equal to the atmospheric pressure p_a. As a result, the Bernoulli theorem implies that the velocity decreases with the applicate.

NOTE.– In energetic terms, since the potential pressure energy does not vary, this means that an exchange can along take place between the kinetic energy of a fluid particle and its potential gravitational energy.

The principle of continuity therefore implies that since the flow is conserved, and the fluid is incompressible, S is inversely proportional to the velocity. As a result, $S(z)$ will be an increasing function. As stated in the text, S should be "theoretically infinite" when the velocity is equal to zero.

1.2) Let us write out the Bernoulli theorem between a point A located below the piston and a point B located on the same current line in the jet, with an applicate value of z_B:

$$p_T = \rho\frac{V^2(z_A)}{2} + p_A + \rho g z_A = \rho\frac{V^2(z_B)}{2} + p_a + \rho g z_B \qquad [4.48]$$

The origin of the cotes is chosen on the superior side of the piston, and the velocity is considered negligible compared with the velocity of a point of the jet:

$$z_A = -a;\ V^2(z_A) << V^2(z_B) \qquad [4.49]$$

The pressure P_A is the sum of the atmospheric pressure and the overpressure caused by the weight mg of the piston acting on a surface $S = \pi\frac{D^2}{4}$:

$$p_A = p_a + \frac{mg}{S} \qquad [4.50]$$

This results in:

$$V(z_A) = \sqrt{\frac{2(p_A - p_a)}{\rho} + 2g(z_A - z_B)} = \sqrt{\frac{2mg}{\rho S} - 2g(z_B + a)} \qquad [4.51]$$

2) The extremity of the jet is "theoretically" reached when the velocity is equal to zero, which happens for:

$$\frac{2mg}{\rho S} - 2g(z_0 + a) = 0 \quad [4.52]$$

$$z_0 = \frac{m}{\rho S} - a \quad [4.53]$$

EXAMPLE 4.7 (Hydraulic circuit with a perfect fluid).–

A circuit is made up of three elements:

– a horizontal branch with a diameter $d = 2\,cm$ and cross-sectional area S_A;

– a vertical branch of cross-sectional area S_A and length $h = 10\,m$;

– a horizontal branch with a progressive change of cross-sectional area (S_B) toward a diameter $D = 2,6\,cm$.

We assume two points A and B: the first located in the horizontal branch of cross-sectional area s and the second one in the horizontal branch of cross-sectional area S. Throughout the entire problem, the pressure in A remains $p_a = 1\,bar$. The circuit is filled with an incompressible and perfect fluid with a density $\rho = 1000\,kg.m^{-3}$.

1) First, the fluid is immobile in the circuit. Give the value of pressure p_{B1} at the point B.

2) Throughout the circuit, the flow rate is equal to 2 l s^{-1}. Give the value of P_{B2} that then becomes the pressure at the point B.

3) Express the relationship between the pressures at points A and B (p_A and p_B), velocity at A (V_A), d and D. Show that there is a flow q_{VC} for which the pressures at A and B are the same. Give the value of this flow q_{VC} and of the velocities then observed at A and B, denoted by V_{AC} and V_{BC}. Express q_{VC} in liters per second.

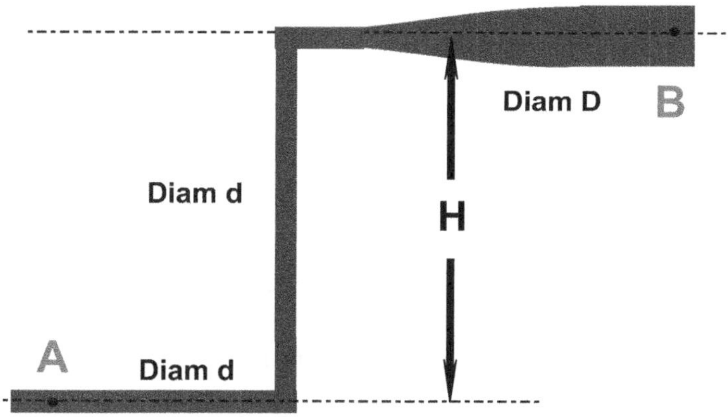

Figure 4.8. *Diagram of the circuit*

Solution:

1) We go "up" by h in the fluid, therefore:

$$p_{B1} = p_A - \rho g h \quad [4.54]$$

$$p_{B1} = 10^5 - 1000 * 9,81 * 10 = 1900\ Pa \quad [4.55]$$

2) The flow rate is known; moreover, it is expressed as a function of the cross-sectional areas and velocities:

$$q_V = 2.10^{-3}\ m^3 s^{-2} = S_A V_A = S_{BV_B} \quad [4.56]$$

$$S_A = 3,14.10^{-4}\ m^2\ ;\ S_B = 5,309.10^{-4}\ m^2 \quad [4.57]$$

$$V_A = \frac{q_V}{S_A} = 6,369\, m.s^{-1} \quad [4.58]$$

$$V_B = \frac{q_V}{S_B} = 3,767\, m.s^{-1} \quad [4.59]$$

The application of the Bernoulli theorem between two points of the same current line belonging to the cross-sectional areas A and B, respectively, results in:

$$\rho \frac{V_A^2}{2} + p_A + \rho g z_A = \rho \frac{V_B^2}{2} + p_B + \rho g z_B \quad [4.60]$$

From which we can obtain the pressure difference as:

$$p_B - p_A = \rho\left(\frac{V_A^2}{2} - \frac{V_B^2}{2}\right) - \rho g\left(z_B - z_A\right) = \rho\left(\frac{V_A^2}{2} - \frac{V_B^2}{2}\right) - \rho g h \qquad [4.61]$$

$$p_B - p_A = -8{,}491.10^4 \ Pa \qquad [4.62]$$

$$p_B = 10^5 - 8{,}491.10^4 \ Pa = 1{,}509.10^4 \ Pa = 0{,}15 bar \qquad [4.63]$$

3) When the pressure at A and B are the same, then:

$$p_B - p_A = 0 = \rho\left(\frac{V_A^2}{2} - \frac{V_B^2}{2}\right) - \rho g h \qquad [4.64]$$

$$\left(\frac{V_A^2}{2} - \frac{V_B^2}{2}\right) = gh \qquad [4.65]$$

$$\frac{V_B}{V_A} = \frac{S_A}{S_B} = \frac{d^2}{D^2} \qquad [4.66]$$

$$\frac{V_{AC}^2}{2}\left(1 - \frac{d^4}{D^4}\right) = gh \qquad [4.67]$$

$$V_{AC}^2 = 301{,}9 \ ; \ V_{AC} = 17{,}38 \, m.s^{-1} \qquad [4.68]$$

$$q_{VC} = S_A \, V_{AC} = 5{,}456.10^{-3} \, m^3 s^{-1} = 5{,}456 \, l.s^{-1} \qquad [4.69]$$

4.3.3. *Draining and filling*

A common category of problems looks at the filling or draining of a reservoir. They are of two types, depending on whether the operation is taking place in a fixed position in relation to the ground, or whether it is mobile.

4.3.3.1. *Fixed reference frame*

In the following problems, the system and the reference frame to which it is tied are fixed in relation to the ground.

EXAMPLE 4.8 (Filling during a competition).–

To quickly fill the reservoir (tank) of a racecar, with a capacity L, a pressurized vessel is used. The air located above the liquid (see Figure) undergoes a pressure of $p > p_a$, where p_a is the atmospheric pressure.

Supply to the car is achieved using a tube T, which is sunk into the bottom of the vessel, while the other end is at atmospheric pressure.

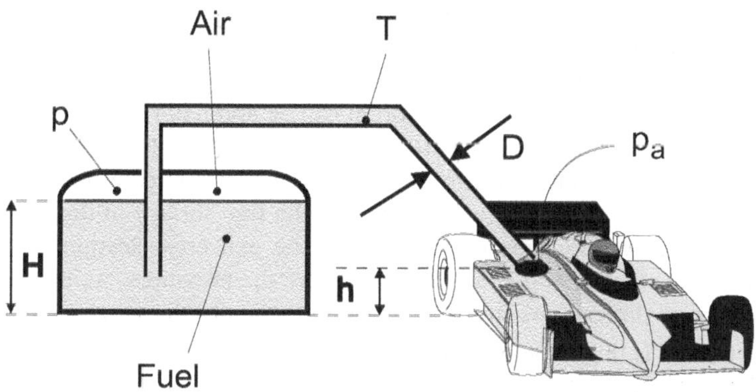

Figure 4.9. *Filling the tank during a race*

1) Before draining of the vessel, calculate the pressure P_0 at the bottom of the vessel. We denote P as the density of the fuel and H as the height of the fuel reserve in the vessel, which is assumed constant.

2) What is the duration t for the filling of the tank? D is the diameter of the tube T. The exit contraction coefficient is equal to 1 (the liquid jet therefore exits as a perfect cylinder).

3) *Numerical application*: We have the following:

$$\rho = 900\,kg.m^{-3};\ H = 1,2\,m;\ H = 1,2\,m;\ h = 60\,cm;$$
$$L = 600\,litres;\quad p_a = 1\,bar;\ p = 10\,bar;\ D = 3\,cm$$

Calculate $H = 1,2\,m$, the filling speed and t_R, the filling time.

For those who watch on television: does the value of t_R seem compatible with modern day racing?

Solution:

1) This is a simple question of fluid statics. We "go down" by H in the reservoir to reach the bottom. The variation in pressure between two horizontal planes separated by the distance H is $\rho g H$. The pressure at the bottom is therefore

$$p_{bottom} = p + \rho g H \qquad [4.70]$$

2) *To know the filling time, the entering flow of fluid into the reservoir must be known*. The cross-sectional area is known, so the velocity of the fluid at the entrance to the reservoir must be determined. Considering the hypotheses (perfect, incompressible fluid, flow that is verified to be stationary, and volume forces that are limited to gravity), the Bernoulli theorem can be applied. However, two relevant points on the same current line must still be chosen. In such a problem, it would be a mistake to choose a point at the bottom, even if the applicate and pressure are known. Instead, we choose a point A located at the free surface of the reservoir and a point B located on the same current line in the exit cross-sectional area of the supply tube. The sides A and B are known, as are the pressures. As the reservoir is large in terms of dimensions, its free surface practically does not change and we have $V_A = 0$. V_B is the velocity that we are looking for. The Bernoulli theorem can be written as:

$$\rho \frac{V_A^2}{2} + p_A + \rho g z_A = \rho \frac{V_B^2}{2} + p_B + \rho g z_A \qquad [4.71]$$

The information provided allows for the following calculations:

We take the sides on an upward vertical axis. The origin is at the level of the bottom of the reservoir (level of the ground):

$$p_A = p; \quad p_B = p_a; \; z_A = H; \; z_B = h; \; V_A = 0 \qquad [4.72]$$

As a result:

$$p + \rho g H = \rho \frac{V_B^2}{2} + p_a + \rho g h \qquad [4.73]$$

$$V_B = \sqrt{2g\left(H - h\right) + 2\frac{p - p_a}{\rho}} \qquad [4.74]$$

The filling flow is written as:

$$q_V = \pi \frac{D^2}{4} V_B \qquad [4.75]$$

And the filling time is the ratio of the volume to fill over the filling flow rate (number of liters of the tank over the number of liters that arrive each second)

$$t_R = \frac{L}{q_V} \qquad [4.76]$$

3) Numerical application:

$$V_B = \sqrt{2g(H-h) + 2\frac{p-p_a}{\rho}} = 44{,}85\, m.s^{-1} \qquad [4.77]$$

$$q_V = \pi \frac{D^2}{4} V_B = 3{,}17 10^{-2}\, m^3.s^{-1} = 31{,}71 l.s^{-1} \qquad [4.78]$$

$$t_R = \frac{L}{q_V} = \frac{600}{31{,}71} = 18{,}9\, s \qquad [4.79]$$

which would not make this filling time particularly advantageous in terms of a competition. Beyond the time value, the problem itself is debatable is terms of the amount of fuel loaded, which would represent a mass of 540 kg.

EXAMPLE 4.9 (Water supply to an island).–

An island is supplied with water by a reservoir with a capacity of $V_{OL} = 1000\, m^3$ located on a hill. The population lives on a plain, which is assumed flat.

The bottom of this reservoir is located at an altitude $H = 30\, m$ above the plain. The reservoir is very flat and has a large surface area on the ground – as such we can neglect the height of the water in the reservoir in relation to the height of the hill: any point in the liquid filling the reservoir is also considered to being at $H = 30\, m$ above the plane. Throughout the whole problem, we assume that water is a perfect fluid. Moreover, the free surface of the water in the reservoir is always equal to the atmospheric pressure p_a. A pipeline with diameter $D = 6\, cm$ goes from the bottom of the reservoir down to the plain. This line supplies the houses of a holidaymaker village. The holidaymakers live in bungalows that are all on the same level. The faucets are all located at a height $h = 1\,m$ above the ground of the plane, and their diameter is $d = 1\,cm$.

1) All the faucets are turned off. What is the pressure p_C that is present on the inside of the faucets?

2) The village is made up of 56 bungalows. It is located in a warm country. Following a communal event, the inhabitants all return home and turn on their faucet, meaning that each house has one faucet open at the same time.

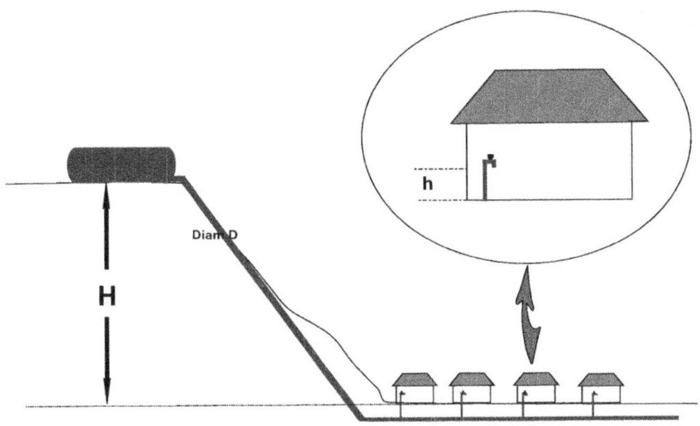

Figure 4.10. *Water supply to a holiday village*

2.1) Calculate the flow rate q_v provided by the faucet.

2.2) During the event, certain dishes of dubious quality were served. Inhabitants from one quarter of the houses were rendered incapacitated and had to be sent to hospital. The faucets in those bungalows stayed on.

a) How much time before the reservoir is fully emptied?

b) The refilling of the reservoir takes a day to organize – will the island have gone any time without water?

3) Everyone recovers well and the reservoir is refilled to 1,000 m^3 of water. Following a bout of terrible weather, the pipeline becomes damaged. It becomes cut at a height $H' = 15\,m$ above the plane. Emptying therefore takes place at this height, through an orifice with the same diameter as the pipe.

The repair will take 4 h. Will the island go without water?

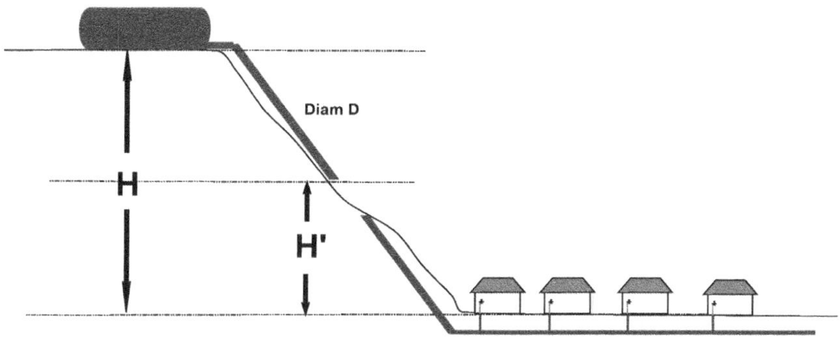

Figure 4.11. *Consequence of a natural disaster*

Solution:

1) Considering the fact we are going down by $\Delta z = H - h$ from a point of the reservoir to the tip of the faucet, a simple application of the fundamental theorem of hydrostatics gives us:

$$p_C = p_a + \rho g \Delta z = p_a + \rho g (H - h) \quad [4.80]$$

2) Health disaster:

2.1) The flow rate of each faucet can be calculated independently.

Let us take the origin of the axis of the Oz altitudes at the level of the ground in the village.

Let us write the Bernoulli theorem between a point A of the reservoir and a point Oz inside the jet coming out of the faucet, both being located on the same current line:

$$\rho \frac{V_A^2}{2} + p_A + \rho g z_A = \rho \frac{V_B^2}{2} + p_B + \rho g z_B \quad [4.81]$$

$$p_C = 10^5 \; p_a + 1000 * 9,81 * 29 = 3,84.10^5 \; Pa = 3,84 \, bar \quad [4.82]$$

The exit velocity of each faucet can be deduced as:

$$\rho \frac{0}{2} + p_a + \rho g H = \rho \frac{V_B^2}{2} + p_a + \rho g h \quad [4.83]$$

$$V_B = \sqrt{2g(H-h)} \qquad [4.84]$$

$$V_B = 23,85\, m.s^{-1} \qquad [4.85]$$

The flow rate of a faucet is written as:

$$q_v = SV_B \qquad [4.86.a]$$

$$\text{with } S = \pi\frac{d^2}{4} = \pi\frac{\left(10^{-2}\right)^2}{4} = 7,85.10^{-5}\, m^2 \qquad [4.86.b]$$

$$q_v = S\, V_B = 7,85.10^{-5} * 23,85 = 1,87.10^{-3}\, m^3.s^{-1} = 112\, litres.mn^{-1} \qquad [4.87]$$

The total flow for the village is:

$$Q_V = 1,87.10^{-3} * 56 = 0,105\, m^3.s^{-1} \qquad [4.88]$$

2.2) The lost flow is:

$$Q_V = q_V \frac{56}{4} = 14 * 1,87.10^{-3} = 2,62.10^{-2}\, m^3.s^{-1} \qquad [4.89]$$

a) The reservoir will empty after the following amount of time:

$$t = \frac{V_{OL}}{Q_V} = \frac{1000}{2,62.10^{-2}} = 3,82.10^4\, s \qquad [4.90]$$

b) t represents $t = 3,82.10^4\, s = 10,6\, hours$ [4.91]

Unfortunately, the village will go without water.

3) Natural disaster:

The draining speed V' with the cote H' can be calculated as in (2.2), using a different value for the difference between the levels $\Delta z' = H - H'$:

$$V' = \sqrt{2g(H-H')} = 17,15\, m.s^{-1} \qquad [4.92]$$

The lost flow is:

$$Q'_V = S'\,V' \qquad [4.93]$$

with $$S' = \pi\frac{D^2}{4} = \pi\frac{\left(6.10^{-2}\right)^2}{4} = 2{,}83.10^{-3}\ m^2 \qquad [4.94]$$

$$Q'_V = S'\,V' = 2{,}83.10^{-3} * 17{,}15 = 4{,}85.10^{-2}\ m^3.s^{-1} \qquad [4.95]$$

which results in a draining time of the reservoir of:

$$t' = \frac{V_{OL}}{Q'_V} = \frac{1000}{4{,}85.10^{-2}} = 2{,}062.10^4\ s = 5{,}73\,hours \qquad [4.96]$$

The repair shall be complete after 4 h. The leak will be repaired in time.

4.3.4. *Mobile reference frame*

In the following problems, the filling takes place in a vessel undergoing inertial movement in relation to the ground.

EXAMPLE 4.10 (In the days of steam engines).–

A device invented in 1862 by the ingenious British engineer Ramsbottom (1814–1897) enabled steam-powered locomotives to be supplied with water without stopping. This device was used in France, on the Paris-Le Havre line, and was located at Léry-Poses.

The system imagined by Ramsbottom was very simple. A channel located between the rails is constantly supplied by a reservoir in such a way that the water level remains constant. The locomotive's tender is equipped with a tube T, or a scoop, which passes through the channel. Water is introduced into the tube T at point O. The water is poured into the tender at the level of point O. The point is located at a very shallow depth under the level of the free surface of the channel. Point O is located at a height h above O. V is the velocity of the locomotive and d is the diameter of the tube T, constant throughout its entire length.

NOTE.– For those who did not live through the happy times of steam engines, let us note that the tender is a carriage of the train, which contains the water reservoir needed to create the steam and carries the fuel.

1) Determine the speed of the water, and then the water flow at O′ as a function of V. ρ is the density of the water.

2) Show that the train must have a minimal velocity V_{min} for the system to function.

3) Numerical application:

3.1) Give the value of the minimal velocity V_{min}.

The velocity of the train being reduced to 70 km.h^{-1} during the filling operation.

3.2) What is the minimum length L of the channel to be able to pour 10 m^3 of water in the tender?

3.3) The real length of the channel was 440 m. Was this enough to place 10 m^3 of water in the tender?

Considering the channel is limited in length, the tube must be lowered and then brought up again, so that it does not hit the extremity of the channel. Knowing that the driver is a highly skilled expert and lowered the tube practically at the start of the channel, how much time does he have to bring the tube back up after the filling of the tender?

We have the following: $\rho = 1000\,kg.m^{-3}$; $d = 20\,cm$; $h = 3\,m$.

Figure 4.12. *Supplying a moving locomotive*

Solution:

We shall use a reference frame linked to the locomotive. Like for any problem of this type we define a vertical upward axis. As we shall see, the origin can be taken at the level of the ground, or, as we shall prefer here, at the level of the axis of tube T. We note that distance from the ground of this axis, denoted by ε, is not given in the numerical applications. We shall see later that this quantity is eliminated during the calculations. In a problem on the mechanics of fluid in a mobile system, we often use a reference system linked to the mobile system. In this case, any fluid that is immobile in relation to the ground behaves as if it were a "bulk" flow, meaning that the flow is "uniform" with a velocity V equal to the velocity of the mobile system.

Let us note that when the mobile system is moving at a constant velocity, which is the case here, the uniform flow is stationary. This therefore fulfills one of the four conditions for the application of the Bernoulli theorem. In the case where the mobile is accelerating, the uniform flow becomes unsteady in the reference frame of the mobile system. Moreover, inertia forces must be considered in the reference frame linked to the mobile system.

As a general rule, the Bernoulli theorem can only be applied between two points of a same current line in two zones where the kinematics are known, or at least simple enough to be deduced.

In a problem like this one, the kinematics of the uniform flow are simple. Another zone where the kinematics are obvious is at the top exit of the tube T. In the location where the fluid exits, for a perfect fluid the flow is uniform.

A common error (which is often made by students) involves placing one of the reference points at the lower extremity (in the channel) of the extraction tube. In this zone, called the "pathological" zone, the flow cannot be deduced simply: it is not uniform and rectilinear. The following question is often asked: the velocity depends on the altitude (since it is constant). Under these conditions, considering the cross-sectional area of the tube is constant, how can the flow be maintained? The answer can be found in the form of the current lines: the flow is supposed to fill the tube T in its superior part. The velocity at the lower part of the tube is clearly higher than that at the top of the tube. Therefore, the flow cannot fill the tube T at its base. The kinetics is therefore complex, the flow is curved and the generating pressure P_G can only be constant over one cross-sectional area. *For this reason, a point in such a zone must never be chosen to apply the Bernoulli theorem.*

1) Let us apply the Bernoulli theorem on a current line between a point A located at the level of the axis of T in a uniform flow with a velocity V upstream of the locomotive, and a point O′ located right at the exit point of the tube T at its summit:

$$\rho\frac{V_A^2}{2}+p_A+\rho g z_A=\rho\frac{V_{O'}^2}{2}+p_{O'}+\rho g z_{O'} \quad [4.97]$$

We replace each element of the equations by the real value, with the unknown one here being $V_{O'}$.

It is important to note that in the uniform flow upstream of the locomotive, on a plane perpendicular to the current lines, or in other words on any vertical plane, $p_G = p + \rho g z$ is constant. Therefore, on a vertical axis between the level of the ground ($z=\varepsilon$) and the level of the axis of T ($z=0$):

$$p_G = p_a + \rho g \varepsilon = p_A + 0 \quad [4.98]$$

$$V_A = V;\; p_{O'} = p_a;\; p_A = p_a + \rho g \varepsilon;\; z_A = 0;\; z_{O'} = h + \varepsilon \quad [4.99]$$

$$\rho\frac{V^2}{2}+p_a+\rho g\varepsilon+0=\rho\frac{V_{O'}^2}{2}+p_a+\rho g(h+\varepsilon) \quad [4.100]$$

The terms p_a and of ρgz are canceled out. There is therefore no need to know p_a and z. As a result:

$$\rho\frac{V^2}{2}+p_a+\rho g\varepsilon+0=\rho\frac{V_{O'}^2}{2}+p_a+\rho g(h+\varepsilon) \quad [4.101]$$

$$V_{O'}=\sqrt{V-2gh} \quad [4.102]$$

This result states that the kinetic energy at the summit of T is equal to the kinetic energy at the level of the horizontal axis of T, increased by the potential energy over the height ε and decreased by the potential energy caused by the height ($h+\varepsilon$).

2) When the fluid arrives at the top of the tube T, its velocity is equal to zero, $V_{O'}=0$. The answer to question 1 provides us with the minimal velocity $V_{\min}$:

$$V_{\min}=\sqrt{2gh} \quad [4.103]$$

This formula is reminiscent of the law governing falling bodies, which is not a coincidence. The kinetic energy of the fluid is fully transformed into potential energy. In a way, this extreme case of use in a zero flow system can be compared to a Pitot tube.

3) Numerical values:

3.1) Value of $V_{\min}$:

$$V_{\min} = \sqrt{2gh} = 7{,}67\,ms^{-1} = 27{,}62\,km\,hr^{-1} \qquad [4.104]$$

3.2)

$$V = 70\,km\,hr^{-1} = 19{,}44\,ms^{-1} \quad ; S = \pi\frac{d^2}{4} = 3{,}14.10^{-2}\,m^2 \qquad [4.105a]$$

$$V_{O'} = \sqrt{V^2 - 2gh} = 17{,}86\,ms^{-1} \qquad [4.105b]$$

$$q_V = S\,V_{\cdot} = 0{,}561\,m^3 s^{-1} \qquad [4.106]$$

This flow must be able to fill 10 m^3. For a train moving at 19,44 ms^{-1}, this results in a distance L of:

$$t = \frac{10}{0{,}561} = 17{,}82\,s \qquad [4.107]$$

$$L = V\,t = 346{,}5\,m \qquad [4.108]$$

3.3) The channel is therefore theoretically quite long, although this does not exempt the controller from having excellent reflexes.

At a speed of 19,44 m/s, the train passes along the channel in 22,63 s. A time of 4,81 s is then left for moving the tube up and down.

EXAMPLE 4.11 (A hardy "Canadair").–

A prototype of an airplane designed to extinguish forest fires uses a water reservoir that is filled with seawater. ρ is the density of this water. To fill the reservoir, the plane has a rod sticking out below it, whose other end is located at the superior part of the reservoir. The lower part of the rod is equipped with a horizontal cross-sectional area. During the filling sequence, this horizontal part is submerged in the water; the airplane flies above the sea at a speed of V ; if V is sufficiently high, the water rises into the rod and fills the reservoir. The rod expels water into the reservoir at a height of H above sea level. The parameter h is the depth at which the bottom of the rod travels through the water; it is also the diameter of the rod at that level.

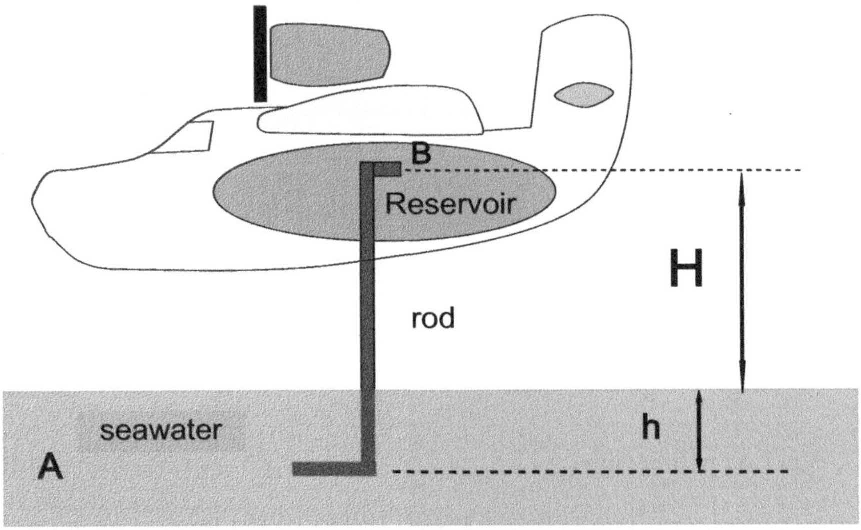

Figure 4.13. *Principle for the filling of a "Canadair"*

1) Give the literal expression for the velocity of the seawater as it exits the upper end of the rod, V_D, as a function of the velocity V of the airplane:

2) We have: $V = 150\,km.hr^{-1}$; $H = 3m$; $d = 5cm$; $\rho = 1000\,kg.m^{-3}$.

The reservoir can contain 10 tons of water.

2.1) Give the numerical values of the velocity and of the volume flow q_V of the water as it exits the rod.

2.2) What is the distance traveled by the airplane over the course of the complete filling of the reservoir?

3) Show that the airplane must be moving at a minimal velocity $V_{\min}$ for the filling to take place. Is this velocity compatible with the plane's own flight?

Solution:

1) This problem is essentially similar to the one involving the filling of the tender on the train. We first draw a vertical upward axis and choose an origin at the

level of the axis of the lower part (submerged) of the tube. By writing the Bernoulli theorem between a point A ahead of the orifice in the water and a point B at the exit of the tube inside the reservoir, we have:

$$\rho\frac{V_A^2}{2}+p_A+\rho g z_A=\rho\frac{V_B^2}{2}+p_B+\rho g z_{B'} \quad [4.109]$$

We replace each element of the equations by its value, with the unknown variable here being $V_{O'}$.

It is important to note that in the uniform flow upstream of the airplane, on any plane perpendicular to the streamlines, or in other words on any vertical plane, $p_G = p+\rho g z$ is constant. Therefore, on a vertical axis, between the level of the ground (here the sea) ($z=\varepsilon$) and the level of the collecting tube ($z=0$), there is:

$$p_G = p_a+\rho g h = p_A + 0 \quad [4.110]$$

$$V_A = V;\; p_B = p_a;\; p_A = p_a+\rho g h;\; z_A = 0;\; z_B = h+H \quad [4.111]$$

$$\rho\frac{V^2}{2}+p_a+\rho g h+0=\rho\frac{V_B^2}{2}+p_a+\rho g\left(h+H\right) \quad [4.112]$$

The terms p_a and $\rho g h$ are canceled out. There is therefore no need to know p_a and h.

As a result, there is:

$$V_B=\sqrt{V^2-2gH} \quad [4.113]$$

2.1) The numerical values of the velocity and flow rate of the water as it exits the rod are:

$$V=41{,}67\,m\,s^{-1};\; V_B=\sqrt{V-2gh}=40{,}95\,m\,s^{-1};\; s=\pi\frac{d^2}{4}=19{,}63\,cm^2 \quad [4.114]$$

$$q_V=s\,V_B=8{,}042.10^{-2}\,m^3\,s^{-1} \quad [4.115]$$

2.2) The filling time t_R is the ratio of the volume of the reservoir, which is 10 m^3, to the filling flow rate, q_V:

$$t_R=\frac{V_{OL}}{q_V}=\frac{10}{8{,}042.10^{-2}}=124{,}3\,s \quad [4.116]$$

which, with a velocity of 41,67 ms^{-1}, results in a distance L of:

$$L = V.t_R = 5179\,m = 5{,}18\,km \qquad [4.117]$$

3) The minimal velocity V_{Lim} of the airplane corresponds to a velocity V_B that is equal to zero, so that the fluid appears at superior extremity of the tube (the length of the cote of the fluid at this level is therefore still $H + h$ and the previous formula remains valid) as:

$$V_B = \sqrt{V_{Lim}^2 - 2gH} = 0 \qquad [4.118]$$

$$V_{Lim} = \sqrt{2gH} = 7{,}67\,ms^{-1} = 27{,}62\,km.hr^{-1} \qquad [4.119]$$

It seems unlikely that an airplane would be able to fly at 27 km.h^{-1}

EXAMPLE 4.12 (A young shepherd fills his water bottle).–

A young shepherd has a brother studying fluid mechanics. Perhaps one day he will be an engineer, but this is not the question here. After reading the paper for one of his brother's tutorials, he gets the idea to fill his water bottle in an interesting way. He gets his hand on a piece of old tubing, of internal diameter $d = 5\,mm$. The vinyl tube is transparent. He takes all the equipment and his bottle into the mountains. He goes to the edge of a torrent, which flows from left to right in the figure. In this simplified problem, we can assimilate the torrent to a flow that is uniform and permanent. The velocity of the water is $U_{torr} = 2{,}5\,m.s^{-1}$.

In this problem, the water is considered an incompressible perfect fluid with a density $\rho = 1000\,kg.m^{-3}$. p_a is the atmospheric pressure.

1) The young shepherd places the tube in the torrent as follows:

The lower part of the tube is horizontal and placed in the torrent at a depth that stays equal to Δz throughout the whole problem (we shall see in the answer that the value of Δz is not important). The tube opens out against the flow, and then bends in a way that links it to the rest of the tube, which is vertical.

The vertical cross-sectional area has a length of $h_1 = 70\,cm$.

What is the height z_1 above the free surface of the torrent reached by the water? Show that the water does not reach the upper extremity of the tube.

2) The shepherd understands that if he walks or runs, the water will climb higher inside the tube. What is the speed U_1 and in what direction in relation to the torrent does the boy have to run for the water to reach a height h_1? To avoid any ambiguity, this direction is defined as left to right or the opposite on the figure. Give U_1 in km.h^{-1}. Must the boy run or can he walk?

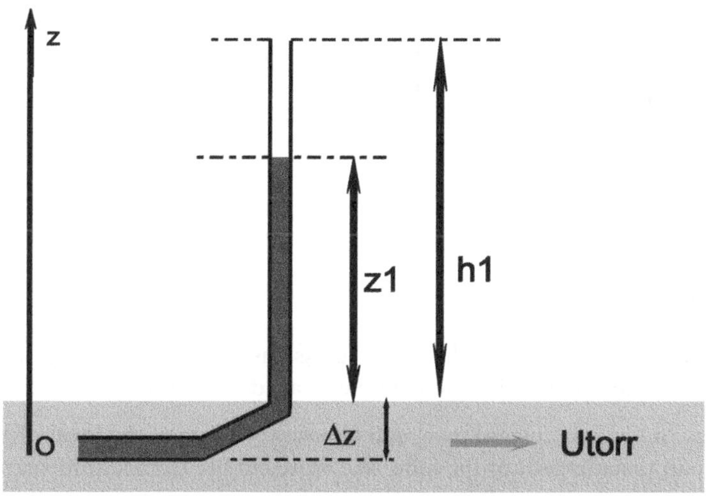

Figure 4.14. *The shepherd runs alongside the torrent*

3) The shepherd then bends the upper part of the tube, which becomes horizontal and goes into the bottle. The vertical part of the tube then has a length $h_1 = 50\,cm$. There is no seal between the tube and the bottle; as a result, the upper extremity of the tube stays at atmospheric pressure.

The little shepherd wants to put 1 l of water in his bottle in less than 2 min. What is the velocity U_2 at which he must run?

4) Was there not a more simple way of filling the bottle? Using this method, how long would it have taken for the liter of water to enter the bottle? The neck of the bottle has a diameter D of 1,5 cm.

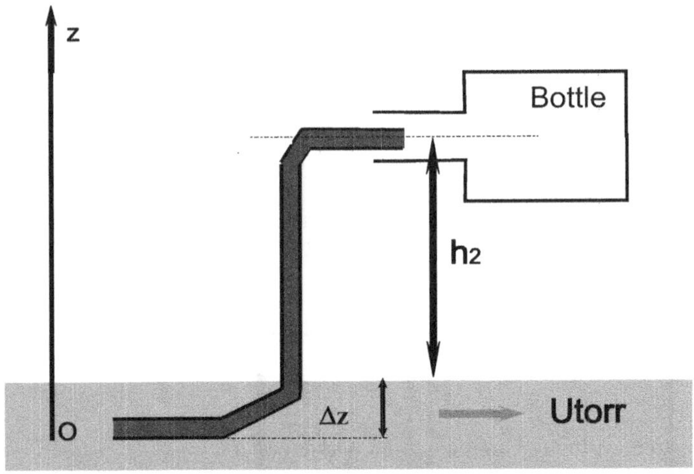

Figure 4.15. *The shepherd fills the bottle*

Solution:

1) This problem belongs to the same "class" as the one involving the tender on the train (Example 4.10) or the one with the "Canadair" (Example 4.11).

We draw a vertical upward axis, and choose its origin at the level of the axis of the lower part (submerged) of the tube. We write out the Bernoulli theorem between a point A very upstream of the orifice in the water, and a point B at the free surface of the liquid in the upper part of the tube. In this question, this level does not reach the top of the tube, and the fluid immobile at B. The flow at A, in the fixed reference frame of the tube, is then uniform, with a velocity of U_{torr}, and its current lines are parallel to the horizontal axis of the tube:

$$\rho\frac{V_A^2}{2}+p_A+\rho g z_A=\rho\frac{V_B^2}{2}+p_B+\rho g z_{B'} \qquad [4.120]$$

We replace each element of the equations by its value, with the unknown here being $z_1=z_B$. We note that in the uniform flow upstream of the tube, on any plane perpendicular to the lines of the current, which here is any vertical plane, $p_G=p+\rho g z$ is constant. Therefore, on a vertical axis, between the level of the axis of the tube ($z=0$) and the level of the free surface of the torrent, we have $z=\Delta z$:

$$p_G=p_a+\rho g\,\Delta z=p_A+0 \qquad [4.121]$$

$$V_A = V; \; p_B = p_a; \; p_A = p_a + \rho g \Delta z; \;\; z_A = 0; \; z_B = \Delta z + H \qquad [4.122]$$

$$\rho \frac{U_{Torr}^2}{2} + p_a + \rho g \Delta z + 0 = \rho \frac{V_B^2}{2} + p_a + \rho g \left(\Delta z + z_1 \right) \qquad [4.123]$$

The terms p_a and ρgz are canceled out. There is therefore no need to know p_a and z.

This means that:

$$\text{v}\, V_B = \sqrt{U_{Torr}^2 - 2gz_1} \qquad [4.124]$$

For $V_B = 0$,

$$U_{Torr}^2 = 2\, g\, \Delta z_1 \qquad [4.125]$$

$$z_1 = \frac{U_{Torr}^2}{2g} = 0{,}318\, m = 31{,}8\, cm \qquad [4.126]$$

2) Let us now write the Bernoulli theorem between a point A in front of the orifice in the water, and a point B_1 at the upper part of the tube, where the fluid has become immobile. The formula established in 1 remains valid:

$$V_B = \sqrt{U_{Torr}^2 - 2gh_1} \qquad [4.127]$$

It is in the interest of the shepherd to maximize the velocity of the fluid in relation to the tube. To do this, he must go against the current. U_{Torr} and U_1 would therefore be added to each other:

$$V_B = \sqrt{\left(U_{Torr} + U_1 \right)^2 - 2gh_1} = 0 \qquad [4.128]$$

$$U_{Torr} + U_1 = \sqrt{2gh_1} = 3{,}7\, ms^{-1} \qquad [4.129]$$

$$U_1 = \sqrt{2gh_1} - U_{Torr} = 1{,}206\, ms^{-1} = 4{,}34\, km\, hr^{-1} \qquad [4.130]$$

The young shepherd is walking at a steady pace against the flow of the torrent.

3) The previous analysis remains valid, V_B is now longer equal to zero. The value of h_1 changes, and becomes $h_2 = 0,5\,m$:

$$V_B = \sqrt{(U_{Torr} + U_1)^2 - 2gh_2} \quad [4.131]$$

$$U_{Torr} + U_1' = \sqrt{V_B^2 + 2gh_2} \quad [4.132]$$

$$q_V = sV_B = \pi\frac{d^2}{4}\sqrt{(U_{Torr} + U_1)^2 - 2gh_2} \quad [4.133]$$

$$s = \pi\frac{d^2}{4} = 1,96.10^{-5}\,mm^2 \;\; ; \;\; q_V = \frac{10^{-3}}{120} = 8,33.10^{-6}\,m^3\,s^{-1} \quad [4.134]$$

$$V_B = 0,425\,m.s^{-1} \quad [4.135]$$

$$U_{Torr} + U_2 = 3,161\,m.s^{-1} \quad [4.136]$$

$$U_2 = 0,66\,m.s^{-1} = 2,37\,km\,hr^{-1} \quad [4.137]$$

4) Yes, the young shepherd can put the bottle directly and horizontally into the water:

$$S = 176.10^{-6}\,m \;\; ; \;\; q_V = SU_{Torr} = 4,4.10^{-4}\,m^3 s^{-1} = 0,44\,litres.s^{-1} \quad [4.138]$$

$$t_R = \frac{10^{-3}}{0,44.10^{-4}} = 2,27\,s \quad [4.139]$$

The filling would then take a little more than 2 s (2.27 s). In any case, this calculation is only one for determining an order of magnitude. Implicitly, the previous calculation assumed that a tube with a current of diameter D penetrated the bottle at a velocity equal to that of the torrent. In reality, the kinematics are likely to be more complicated.

4.3.5. *Time-dependent filling*

Other problems regarding filling or draining are more complicated, as the velocity of the fluid is no longer constant over time.

This velocity depends on the levels of the fluid in the reservoir in question. Resolution of the problem and calculations surrounding the filling or draining times involve a differential equation. There are, however, many similarities in how these problems are resolved.

EXAMPLE 4.13 (In Cleopatra's time).–

Egyptologists have theorized that the architects of Ancient Egypt used ingenious ways to slowly lower large stones.

This process is represented in a movie to spectacular effect, where it is used to seal off the tomb where Cleopatra's historical story ends. Such a system is also present during the tragic end of *Radamès* and *Aîda,* main characters of the Verdi opera titled Aïda. Here we shall create a model of such a device. A stone with a height h and a square base a can slide without friction through a square stone chimney, the sides of which are roughly equal to a. As the diagram shows, the stone lies on a bed of sand. In the initial moment, the sand is allowed to pour out through a hole with a diameter of d. There is: $d \ll a$. H_0 is the initial height of the bottom of the stone. The densities of the stone and sand are ρ_P and ρ_S, respectively. The sand is considered a perfect and incompressible fluid. Some hypotheses used here are perhaps less than realistic, notably the absence of fluid–solid or solid–solid friction. The model used here is a simplification of reality.

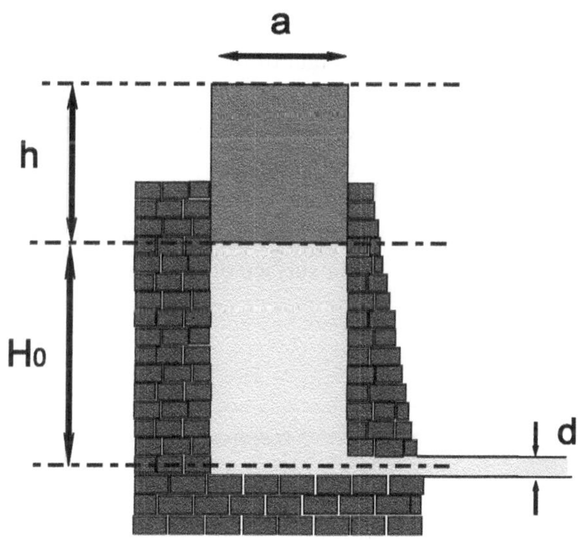

Figure 4.16. *In the time of pyramids: clever engineering*

1) Calculate the static pressure within the sand upon contact with the stone, p_S.

2) Calculate the exit velocity V of the sand as a function of the height of the bottom of the stone, z, counted here from above the applicate 0.

3) What is the equation for the height z?

Deduce the movement of the stone $z(t)$.

4) What is the lowering time of the stone t_D?

5) Numerical application. Calculate t_D:

$\rho_P = 2700\,kg.m^{-3}$; $\rho_S = 2500\,kg.m^{-3}$; $H_0 = 15m$; $a = 3m$; $h = 2m$; $d = 20\,cm$; the jet of sand has a contraction coefficient of $C_c = 1$.

Solution:

This classic problem is part of the category of the draining of receptacles or reservoirs. It is characterized by the application of a pressure higher than the atmospheric pressure to the higher part.

1) A pressure is the ratio of a force to a surface.

The force $\vec{F}$ that is applied to the high part of the sand, meaning the lower part of the stone, is applied over a square of cote a, with an area of a^2.

This force $\vec{F}$ is the result of the weight of the stone (produced by its volume, its density and by g), to which is added the force $\vec{F}'$ caused by atmospheric pressure p_a applied to the upper cross-sectional area of the stone (which the stone "transmits" to its base, and therefore to the sand).

This gives us the calculation of the pressure p_S, which is the ratio of the sum of these two forces to the area of the lower surface of the stone:

$$F = \rho_P\, a^2 hg;\;\; F' = p_a a^2 \qquad [4.140]$$

$$p_S = \frac{F + F'}{a^2} = \frac{\rho_P a^2 hg + p_a a^2}{a^2} \qquad [4.141]$$

$$p_S = \rho_P\, hg + p_a \qquad [4.142]$$

2) We mark out the space with a vertical upward axis and choose the origin at level of the hole with a diameter d through which the sand pours out.

The two principles are used: principle of continuity, (conservation of the flow) and the fundamental principle of dynamics (in the form of the Bernoulli theorem).

Principle of continuity: the flow of sand in the upper part is equal to the draining flow. V_S is the velocity of the sand in its upper part and V is the exit velocity at the hole. The principle of continuity means that the flow is conserved between the superior cross-section, the area of which is $S = a^2$ and the jet exiting the hole, the area of which is $s = C_C \pi \frac{d^2}{4}$. Meaning:

$$q_V = S V_S = s V \qquad [4.143]$$

This relation will be used later on.

In this question, it will only be used to compare V_S and V:

$$\frac{V_S}{V} = \frac{s}{S} = \frac{\pi d^2}{4a^2} < 1; \left(\frac{V_S}{V}\right)^2 \ll 1 \qquad [4.144]$$

For later, the following is defined:

$$\alpha = \frac{s}{S} = \frac{\pi d^2}{4a^2} \qquad [4.145]$$

Fundamental principle of dynamics. We write out the Bernoulli theorem on the same current line between two zones of interest:

– a point on the upper surface of the sand, S;

– a point in the jet. As this jet has a small diameter, we can assume that pressure within it is constant and, for reasons of continuity, equal to the atmospheric pressure p_a:

$$\rho \frac{V_S^2}{2} + p_S + \rho_S g z = \rho \frac{V^2}{2} + p_a + \rho_S \, g(0) \qquad [4.146]$$

Knowing that $\left(\frac{V_S}{V}\right)^2 << 1$, V can be deduced as:

$$V = \sqrt{2gz + \frac{2(p_s - p_a)}{\rho_S}} = \sqrt{2gz + \frac{2\rho_P hg}{\rho_S}} \tag{4.147}$$

$$V = \sqrt{2g\left(z + \frac{\rho_P h}{\rho_S}\right)} \tag{4.148}$$

3) We write out the differential equation that governs the cote $z\ (t)$ of the height of the sand. It is important to note that, even if the ratio α is *very low*, V_S is *not equal to zero:*

$$V_S = \alpha V = \alpha\sqrt{2g\left(z + \frac{\rho_P h}{\rho_S}\right)} \tag{4.149}$$

Moreover, this velocity V_S is equal at least to the derivative of the cote z(t) in relation to time. It should be noted that z decreases with time and that V_S is a positive parameter, and thus we can write:

$$V_S = -\frac{dz}{dt} \tag{4.150}$$

By combining the previous expressions:

$$\frac{dz}{dt} = -V_S = -\alpha V = -\alpha\sqrt{2g\left(z + \frac{\rho_P h}{\rho_S}\right)} \tag{4.151}$$

Therefore: $$\frac{dz}{dt} = -\alpha\sqrt{2g\left(z + \frac{\rho_P h}{\rho_S}\right)} \tag{4.152}$$

which is the differential equation we are looking for. This first-order equation requires knowledge of a condition at the limits. We know the cote from the top of the sand at the initial moment:

$$t = 0 \ ; \ z = H_0 \tag{4.153}$$

Practical tip. The equation contains rather complex terms. It is recommended to simplify its expression by introducing the coefficients A and B:

$$\frac{dz}{dt} = -\alpha\sqrt{Az + B} \qquad [4.154]$$

With:

$$A = 2g; \; B = 2g\frac{\rho_P h}{\rho_S} \qquad [4.155]$$

Faced with this type of expression, it is easier to express t as a function of z. We therefore group the terms that contain z on one side and the terms with t on the other side:

$$\frac{dz}{\sqrt{Az + B}} = -dt \qquad [4.156]$$

The term on the left-hand side of the equation then appears as the differential of an equation:

$$\frac{dz}{\sqrt{Az + B}} = d\left[\frac{2}{A}\sqrt{Az + B}\right] \qquad [4.157]$$

We replace dz in the differential equation with the expression found in [4.156]:

$$d\left[\frac{2}{A}\sqrt{Az + B}\right] = -\alpha\, dt \qquad [4.158]$$

If we integrate the two terms, meaning we pass from the differentials to the functions and exchange the places of the terms in the left- and right-hand sides, we obtain:

$$\alpha t = -\frac{2}{A}\sqrt{Az + B} + Cte \qquad [4.159]$$

We apply the condition at the limits to obtain the value of the constant *Cte* Cnst, which solves the problem:

$$t = 0;\ z = H_0;\ Cte = \frac{2}{A}(AH_0 + B)^{\frac{1}{2}} \tag{4.160}$$

$$\alpha t = -\frac{2}{A}\sqrt{Az + B} + \frac{2}{A}\sqrt{AH_0 + B} \tag{4.161}$$

Finally, by dividing both sides by α:

$$t = -\frac{2}{A\alpha}\sqrt{Az + B} + \frac{2}{A\alpha}\sqrt{AH_0 + B} \tag{4.162}$$

4) The lowering time of the stone, t_D, corresponds to the moment where z is canceled out.

It is expressed as:

$$t_D = -\frac{2}{A\alpha}\sqrt{B} + \frac{2}{A\alpha}\sqrt{AH_0 + B} \tag{4.163}$$

5) We have:

$$A = 19{,}62\,SI;\ B = 42{,}38\,SI;\ \alpha = 3{,}49.10^{-3};\ \frac{2}{\alpha A} = 29{,}21 \tag{4.164}$$

$$t_D = -\frac{2}{A\alpha}\sqrt{B} + \frac{2}{A\alpha}\sqrt{AH_0 + B} \tag{4.165}$$

$$t_D = 346\,s = 5{,}8\,mn \tag{4.166}$$

EXAMPLE 4.14 (Submersion of a submarine).–

A submarine has an external volume $V_{Sub} = 16\,000\,m^3$. We can roughly model the submarine in the shape of a right-angle parallelepiped with a length $L = 120\,m$, width $l = 14\,m$ and height $H = 9{,}5\,m$. The value of H is not used in the problem. However, it can be used to verify some of the results.

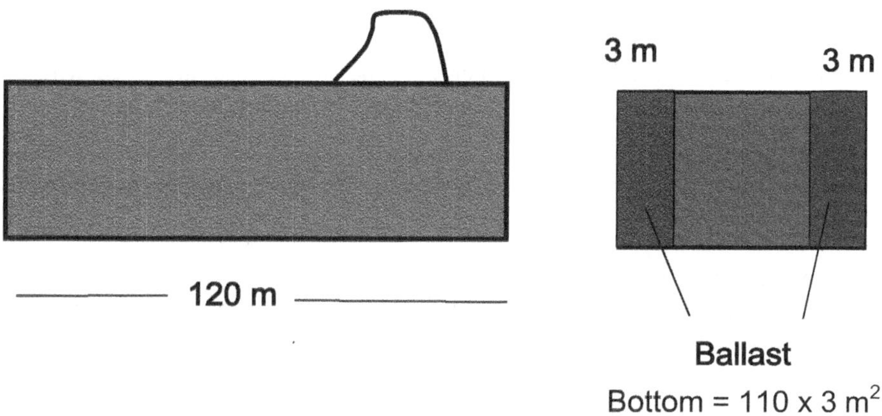

Figure 4.17. *The submarine*

This submarine is made up of two identical ballasts located on the side of the hull. Each has a flat bottom with a length $L_B = 110\,m$ and width $l_B = 3\,m$. Depending on how the ballasts are filled, the submarine can float or it sinks.

The mass of the submarine is $M = 1200\,t$ when the ballasts are empty. The density of water is $\rho = 1000\ kg.m^{-3}$. The atmospheric pressure is $p_a = 1\,bar$. The water is considered a perfect fluid throughout the entire problem.

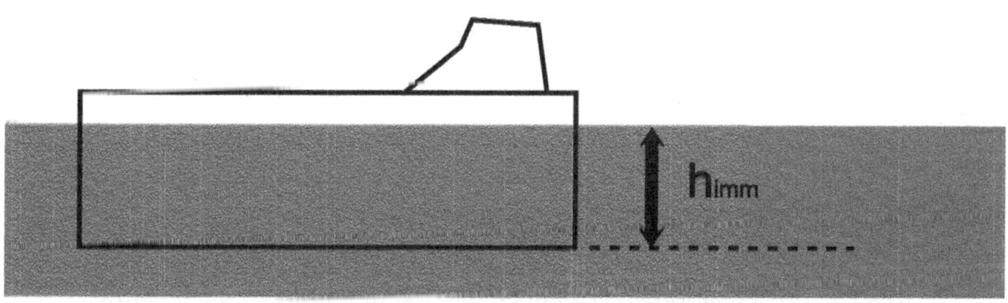

Figure 4.18. *The submerging submarine*

1) The ballasts are empty. The submarine floats. The height h_{imm} of the submerged part of the hull is called the draft when the ship is floating. What is this draft here?

2) What minimal mass of water M_{Ball} must be introduced into each ballast for the submarine to become completely submerged? The two ballasts are filled with the same volume. What is then the height of the volume of water h in each ballast?

M_{Ball} is the amount of water that in introduced in practice. It creates a different balance. The submarine can then go up or down in the water using its propellers and fins.

3) The submarine is at port. Accidentally, the ballasts are filled to a height h, calculated in question 2 for complete submersion. Fortunately, the ship is properly secured and its depth of submersion remains h_{imm} calculated in question 1.

The ballasts are then emptied. For this, a pressure if h_{imm} is applied through compressed air, and is maintained over time. Each of the ballasts is then drained through an orifice with a diameter of 10 cm located at its bottom. The parameter $x(t)$ is the height of the water in a ballast at a time t during draining. The function x(t) is clearly the same for each of the two ballasts.

3.1) What is the static pressure p_{OR} in the horizontal plane passing through the orifices of the ballasts?

3.2) Noting that the speed of descent from the free surface V_S in each ballast is very much lower than V_D, the draining speed at the exit orifice, express V_D as a function of $x(t)$.

3.3) Write the relation between the draining speed at the exit orifice, V_D, and the derivative $\frac{dx}{dt}$.

3.4) Write out the differential equation solved by $x(t)$.

3.5) Resolve this equation and calculate the draining time for each ballast.

Solution:

1) The submarine is in equilibrium under the influence of gravity and of the Archimedes principle:

$$Mg = \rho lLh_{imm}g; \quad h_{imm} = \frac{M}{\rho lL} \qquad [4.167]$$

$$h = 7{,}14\,m \qquad [4.168]$$

2) Considering a submerged volume to that of the submarine V_{sub}, the new equilibrium can be written as:

$$(M + 2\,M_{Ball})\,g = \rho V_{sub}\,g \qquad [4.169]$$

$$M_{Ball} = \frac{\rho V_{sub} - M}{2} \;;\; M_{Ball} = 2.10^6\,kg = 2000\,t \qquad [4.170]$$

$$h = \frac{M_{Ball}}{\rho\, l_B\, L_B} = 6{,}06\,m \qquad [4.171]$$

4.4. Draining of the ballasts

3.1) What is the static pressure p_{or} in the horizontal plane passing through the orifices of the ballasts?

In the water of the port, the system is "lowered" from the free surface to a depth of h_{imm}. The increase in pressure is therefore $\rho\, g\, h_{sub}$, and

$$p_{or} = p_a + \rho g h_{sub} \qquad [4.172]$$

$$p_{or} = 1{,}7.10^5\,Pa = 1{,}7\,bar \qquad [4.173]$$

3.2) First note that the equation of continuity allows us to write out the flows of water in each ballast between the free surface, with an area of $l_B\, L_B$ t, and the orifice with an area of $\pi \frac{d^2}{4}$ as equal:

$$l_B\, L_B\, V_{surf} = \pi \frac{d^2}{4} V_D \;;\; V_{surf} = \pi \frac{d^2}{4 l_B\, L_D} V_D \qquad [4.174]$$

$$V_{surf} = 2{,}38.10^{-5}\,V_D \;;\; V_{surf}^2 = 5{,}7.10^{-10}\,V_D^2 \qquad [4.175]$$

We mark the applicates on an upward vertical axis and place the origin at the bottom of the ballast. We apply the Bernoulli theorem along the same streamline (which has to exist) between a point of the free surface of the ballast (whose applicate is x, velocity is V_{surf} and is very low and pressure is p_{sup}) and a point located in the orifice, at the bottom and at the exit point of the ballast (whose cote is 0, velocity is V_D and pressure is p_{or}):

$$\rho\frac{V_{surf}^2}{2}+p_{\sup}+\rho g x=\rho\frac{V_D^2}{2}+p_{orp}+\rho g(0) \qquad [4.176]$$

It is important to note that $V_{surf}^2 << V_D^2$:

$$V_D=\sqrt{2\left(\frac{p_{\sup}-p_{orp}}{\rho}+gx\right)} \qquad [4.177]$$

3.3) The continuity equation allows us to write the flow in each ballast between the free surface, with an area of $l_B\,L_B$ and the orifice, with an area of $\pi\frac{d^2}{4}$ as equal:

$$V_{surf}=\pi\frac{d^2}{4l_B\,L_B}V_D \qquad [4.178]$$

While very low, V_{surf} is not equal to zero. Moreover, the velocity V_{surf} is the speed of "descent" from the free surface. As the cote of the surface, x, is counted on an upward axis, the derivative of x is negative. VD is a norm of the velocity, and therefore positive, so there is:

$$V_{surf}=-\frac{dx}{dt} \qquad [4.179]$$

$$\frac{dx}{dt}=-\pi\frac{d^2}{4l_B\,L_B}V_D \qquad [4.180]$$

3.4) Considering the expression of V_D found previously, which is a function of x, we obtain the desired equation:

$$\frac{dx}{dt}=-\pi\frac{d^2}{4l_B\,L_B}V_D=-\pi\frac{d^2}{4l_B\,L_B}\sqrt{2\left(\frac{p_{\sup}-p_{orp}}{\rho}+gx\right)} \qquad [4.181]$$

As always with such an expression, the writing should be simplified by introducing two constants A and B, which can be calculated from the information given in the problem:

$$\frac{dx}{dt} = -\sqrt{Ax + B} \tag{4.182}$$

$$A = \pi\frac{d^2}{4l_B L_B}\sqrt{2g} = 1{,}11.10^{-8}\,SI \;\; ; \;\; B = \pi\frac{d^2}{4l_B L_B}\sqrt{2\left(\frac{p_{\text{sup}} - p_{orp}}{\rho}\right)} = 9{,}4.10^{-7}\,SI \tag{4.183}$$

3.5) The differential equation has a relatively common form in problems relating to filling and draining. It is more useful to resolve it by expressing t as a function of x:

$$\frac{dx}{\sqrt{Ax+B}} = -dt \tag{4.184}$$

$$\frac{2}{A}\sqrt{Ax+B} = -t + C \tag{4.185}$$

Considering the condition at the limits $t = 0\ x = h_{imm}$:

$$\frac{2}{A}\sqrt{Ax+B} = -t + \frac{2}{A}\sqrt{Ah+B} \tag{4.186}$$

The draining is finished when x is equal to zero. The time t_D is therefore:

$$t_D = \frac{2}{A}\left(\sqrt{Ah+B} - \sqrt{B}\right) = 6142\,s = 1\,h\,42\,mn\,22\,s \tag{4.187}$$

4.5. Synthetic problems

Finally, we will finish with some synthetic problems that call on all of the techniques seen above.

EXAMPLE 4.15 (Artesian wells).–

A cave contains water. The rock is porous between the cave and a point A located in the valley. This way, the water can pass through the ground.

All of the required parameters are defined in the figure. We can consider the cave as a reservoir where the fluid is in a resting state and that the porous zone is made up of a large number of channels that link the cave and the ground. The free surface of the reservoir is at an altitude H above the ground.

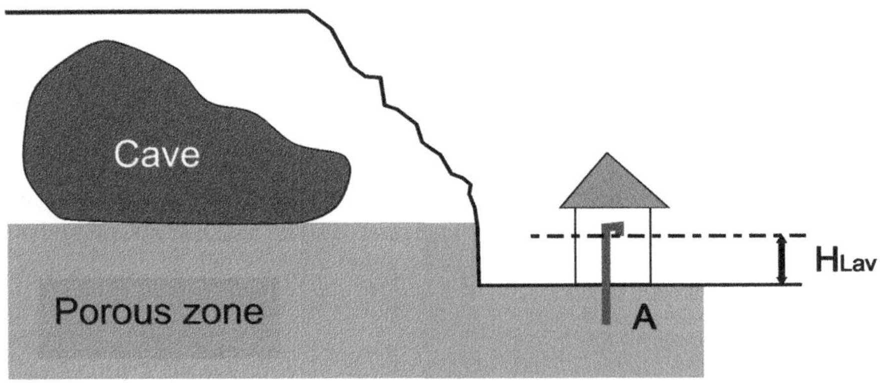

Figure 4.19. *An artesian well*

1) To what height h in relation to the ground will the water spurt up?

2) The flow in A is captured using a tube with a diameter d and drinkable (or at least we assume) is obtained. What is the maximum volume flow q_V of water that be obtained in a faucet on the third floor located at an altitude of H_{Fau} in relation to A?

3) Numerical application. We have $H = 30\,m$; $H_{Fau} = 9\,m$; $\rho_{water} = 1000\,kg.m^3$; $d = 14\,mm$. Give the values of h (question 1) and q_V (question 2)? Does the value obtained for q_V seem realistic? Justify the answer qualitatively.

Solution:

This problem is synthetic in that mixes questions relating to the determination of velocities and of cotes (draining and jets).

1) We create a frame for the space using an upward vertical axis Oz. The origin O is fixed at the ground level.

A jet is a "free" flow, meaning that it touches no solid surfaces. For perfect fluid jet, the Bernoulli theorem can be applied.

For such a flow, the total pressure $p_T = \rho\frac{V^2}{2} + p + \rho gz = \rho\frac{V^2}{2} + p_G$ is constant.

In a free jet flowing into the atmosphere, for reasons of continuity, the pressure is constant and equal to atmospheric pressure.

This can be shown rigorously for a vertical jet, which can be considered as a vertical flow (which is an approximation, as the jet is necessarily expanding). In any plane that is perpendicular to the lines of current, in this case horizontal, p_G is constant, so $p_G = p + \rho gz = Cnst.$ On a horizontal plane, z is constant; therefore, *in the jet*, the pressure p is constant. At the interface between the jet and the air, the pressure is continuous. p is therefore equal to atmospheric pressure p_a in both the air and liquid. This is the case no matter the altitude z :

$$\rho\frac{V^2}{2} + p + \rho gz = \rho\frac{V^2}{2} + p_a + \rho gz \qquad [4.188]$$

As a result, V and z vary in opposite directions. There is therefore an altitude at which the jet has a velocity of zero. We must look at the limits of the reasoning: by all rigor, the conservation of the flow means that the cross-sectional area increases as the velocity decreases. The lines of the current are therefore not rigorously parallel. Moreover, a velocity of zero tends toward an infinite cross-sectional area at the limit. For a calculation of orders of magnitude, which includes any use of the Bernoulli theorem, this reasoning is sufficient. In reality (a water jet in a park, for example), the jet does not keep its integrity, becomes unstable and breaks into drops. By applying the Bernoulli theorem on a current line between the point A of the free surface of the cave and the point B at the top of the jet, we obtain:

$$\rho\frac{V_A^2}{2} + p_A + \rho gz_A = \rho\frac{V_B^2}{2} + p_B + \rho gz_B \qquad [4.189]$$

By applying the values given with the problem:

$$\rho(0) + p_a + \rho gH = \rho(0) + p_a + \rho gh \qquad [4.190]$$

The following is immediately apparent:

$$h = H \qquad [4.191]$$

This can be interpreted easily: the fluid in the reservoir has a potential volume energy of ρgH. This energy is transformed into kinetic energy in the jet. At the top of the jet, any kinetic energy is returned to its potential form. The "kinetic" and "gravitational" energies are exchanged, as the pressure energy is constant along the whole jet submerged in the air, which is at atmospheric pressure.

2) We apply the Bernoulli on a current line between the point A of the free surface in the cave and the point B at the top of the jet.

At the altitude h_{Lav}, the velocity V is such that:

$$\rho(0) + p_a + \rho gH = \rho\frac{V^2}{2} + p_a + \rho gh_{LAV} \qquad [4.192]$$

Therefore:

$$V = \sqrt{2g(H - h_{LAV})} \qquad [4.193]$$

The flow is therefore:

$$q_V = SV = S\sqrt{2g(H - h_{LAV})} \qquad [4.194]$$

where $S = \pi\frac{d^2}{4}$ is the cross-sectional area of the tube.

3) Numerical application. $h = H = 30 m$:

$$q_V = SV = S\sqrt{2g(H - h_{LAV})} \qquad [4.195]$$

$$S = \pi\frac{d^2}{4} = 1{,}54 10^{-4}\, m^2 \ ; \ V = 14{,}69\, ms^{-1} \qquad [4.196]$$

$$q_V = 2{,}26 10^{-3}\, m^3 s^{-1} = 2{,}26\ litres.s^{-1} \qquad [4.197]$$

NOTE.– This flow comes out as $q_V = 136\ liters.mn^{-1}$, which can quite understandably be considered a bit excessive (a household faucet has a flow of

10 $liters.mn^{-1}$). This is an inherent problem involving the blind application of the Bernoulli theorem. Under real conditions, head losses must also be considered.

EXAMPLE 4.16 (Injection problems).–

Here we look at the different phases involved as a nurse performs the injection of a medication into a patient.

Part A. – During the injection:

A nurse must inject a patient with a quantity of medication equal to $8\,cm^3$. Despite its high price, this medication is made up mainly of distilled water, and as such has a density $\rho = 1000\,kg.m^{-3}$. To do this, she inserts the needle of the syringe into the patient at the relevant location. We assume that inside the patient the pressure is equal to atmospheric pressure P_a present outside the patient. The internal component of the syringe is a cylinder of diameter $D = 2\,cm$. The internal diameter of the needle where the liquid exits is equal to $d = 0,5\,mm$.

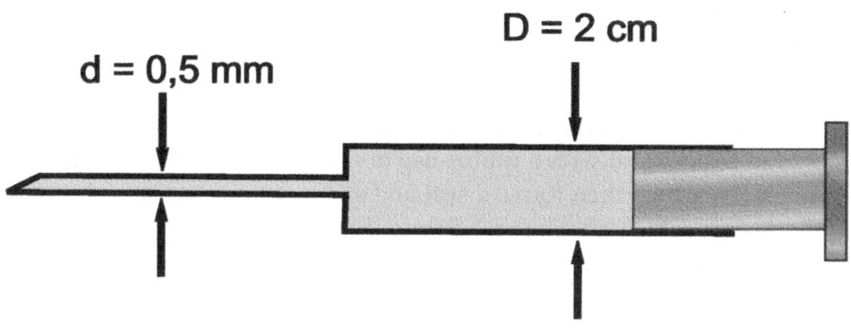

Figure 4.20. *Diagram of the syringe*

1) The nurse wants to inject the $8\,cm^3$ of medication in $10\,s$ at a constant velocity V.

1.1) What is the instantaneous flow rate q_v inside the patient? What is the velocity V ?

1.2) What pressure must be applied by the piston on the liquid in the syringe?

1.3) What force F must be applied by the nurse's finger onto the syringe? *Note that the effect of this force is added to the effect of the external atmospheric pressure.*

2) Because of a high workload, the nurse decides to accelerate the rhythm and empty her syringe in 1 s. What force F' must she then apply to the syringe? Express this force in kilograms-force (kgf). Will she have the strength to accelerate her task?

Unfortunately, it is often necessary to re-state that a kilogram-force is the unit of force, the intensity of which is equal to the weight of a mass of 1 kg.

Part B. – Before injection:

The nurse has had to inject a patient with an amount of medication equal to 8 cm^3, which, while expensive, is made up mainly of distilled water, giving it a density ρ equal to 1,000 kg m^{-3}. Moreover, this liquid is considered a perfect fluid. The internal aspect of the syringe is a cylinder with a diameter $D = 2cm$. The internal diameter of the needle where the liquid exits is $d = 0,5\,mm$.

Part B. – Problem 1. Filling the syringe:

Before injection, the syringe must be filled. The products to be in injected are often stored in a glass vial with a rubber cap at the top. To fill the syringe, the nurse pierces the cover, which then forms a seal and sucks the product into the syringe by pulling the piston, which acquires a velocity V_P. All nurses know that if they pull too quickly on the piston, the product is not aspirated. We shall explain this phenomenon here.

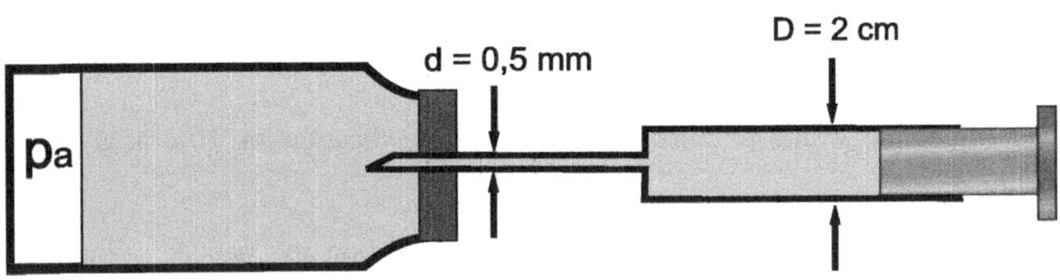

Figure 4.21. *Filling the syringe*

For this, we use the following hypotheses:

– the vial is considered a reservoir mainly containing an immobile liquid;

– atmospheric pressure governs the liquid.

We assume that the velocity of the liquid at the level of the piston is equal to the velocity of the piston.

1) Deduce the velocity inside the needle of the syringe.

2) Give the expression for the pressure p_{nee} in the needle.

3) Show that from a certain value of V_P, a pocket of water vapor appears in the needle and blocks aspiration. The parameter p_S represents the tension of the saturating vapor of the water at ambient temperature.

4) Numerical application.

At room temperature, the tension of the saturating water vapor is equal to $20\,mm$ of mercury.

4.1) At what value of V_P does the aspiration process start to be blocked?

4.2) What is then the minimal time τ required for 8 cm^3 of product to enter the syringe?

Part B. – Problem 2. Eliminating air:

Just before injection, the nurse does the well-known "squirt" gesture to evacuate any air that could be present in the needle.

As we know, any air present in a blood vessel causes coagulation, the consequences of which could from passing pains to a lethal embolism. For this reason, she points the needle upward and pushes the piston with a force of F. L is the distance at that moment between the top of the piston and the extremity of the needle. A jet escapes from the needle. This jet climbs to an altitude of h, where its velocity becomes zero.

1) Show that the velocity V_L of the liquid decreases from the moment it leaves the needle.

2) What is the height above the needle reached by the jet?

3) A.N. If $F = 2,6\,N$ and $L = 10\,cm$, find the value of h.

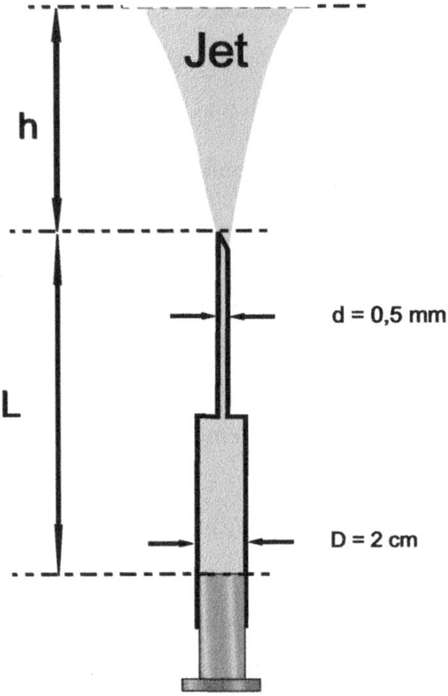

Figure 4.22. *Evacuating the air*

Solution:

A) During injection

1) Thoughts and preliminary calculations.

The forces applied to the piston of the syringe cause a pressure to be applied to the fluid in contact with the piston that is higher than the atmospheric pressure. As the system is horizontal, gravity plays no role and the changes in energy take place between the potential pressure energy and kinetic energy.

We can therefore apply the Bernoulli theorem between a point A in contact with the piston inside the syringe and a point B located on the same line of current just at the exit of the needle.

1.1) S is the internal cross-sectional area in the body of the syringe and s is the exit cross-sectional area of the needle. Furthermore, V_p is the velocity of the piston and V is the velocity of the fluid as it exits the needle. Over the $10\,s$ of injection – which is assumed at a constant speed – the flow rate is:

$$q_V = \frac{8.10^{-6}}{10} = 8.10^{-7}\, m^3.s^{-1} \qquad [4.198]$$

$$q_V = sV = \pi\frac{d^2}{4}V \;\; ; \;\; s = 1{,}96.10^{-7}\, m^2 \qquad [4.199]$$

$$V = 4{,}08\, m.s^{-1} \qquad [4.200]$$

The principle of continuity allows us to write:

$q_V = sV = S\,V_P$ where V_P is the velocity of the piston.

Therefore:

$$\frac{s}{S} = \left(\frac{d}{D}\right)^2 << 1 \;\; ; \;\; V_P << V \qquad [4.201]$$

We choose a point A in the fluid in contact with the piston and a point B in the jet at the exit point of the needle, on the same current line. The Bernoulli theorem is written as:

$$\rho\frac{V_A^2}{2} + p_A + \rho g z_A = \rho\frac{V_B^2}{2} + p_B + \rho g z_B \qquad [4.202]$$

$$p_A = p_P;\; p_B = p_a;\; z_A = z_B;\; V_A = V_P;\; V_B - V \qquad [4.203]$$

$$\rho\frac{V_P^2}{2} + p_P = \rho\frac{V^2}{2} + p_a;\; V_P^2 << V^2 \qquad [4.204]$$

$$p_P = \rho\frac{V^2}{2} - \rho\frac{V_P^2}{2} + p_a \approx \rho\frac{V^2}{2} + p_a \qquad [4.205]$$

1.2) The pressure on the piston is the sum of the atmospheric pressure and the additional pressure caused by the force $\vec{F}$. As a pressure is the ratio of a force to an area S, and considering [4.205]:

$$p_P = p_a + \frac{F}{S} = p_a + \frac{4F}{\pi D^2} = \rho\frac{V^2}{2} + p_a \qquad [4.206]$$

1.3) We can deduce V and F as:

$$V = \sqrt{\frac{8F}{\pi\rho D^2}} \qquad [4.207]$$

$$F = V^2 \frac{\pi\rho D^2}{8} = 2,62\,N \qquad [4.208]$$

2) Accelerate injection

Draining is to take place over a period of time that is 10 times smaller. The mass flow rate will therefore be 10 times higher, as will the velocity V. The force V is proportional to V^2. Therefore, it will be multiplied by 100. $F = 262\,N$. To translate this force into kilogram-force, it must be divided by g. This results in $\frac{262}{9,81} = 26,7\,kgf$. A force of 26,7 kg is undoubtedly excessive for the thumb of a normal nurse.

B) Before injection

Problem 1. Filling the syringe:

1) In this type of question, it is important to work out what we know and what we are looking for to establish a real resolution "strategy". We know the pressure in the reservoir where the velocity is equal to zero and we are looking for the pressure inside the needle. For this, we need to know the velocity within the needle. We know the velocity at the level of the piston, but not the pressure at the level of this piston (which is clearly a depression compared to p_a). The flow must therefore be calculated from the velocity at the piston, the velocity deduced from the flow (continuity principle) and the Bernoulli theorem must be used (fundamental principle of dynamics) to link the pressure of the immobile fluid in the reservoir with the fluid in the needle.

The value of the flow rate is:

$$q_V = \frac{810^{-6}}{10} = 8.10^{-7}\,m^3.s^{-1}$$

The continuity principle allows us to link the respective velocities in the needle and at the piston:

$$q_V = s\,V_{needle} = S_P\,V_P = \pi\frac{d^2}{4}V_{needle} = \pi\frac{D^2}{4}V_P \qquad [4.209]$$

$$V_{needle} = V_P \left(\frac{D}{d} \right)^2 \qquad [4.210]$$

2) We apply the Bernoulli theorem between two points of the same current line, A being in the reservoir ($V = 0$) and B in the needle. The sides are identical, therefore:

$$\rho \frac{V_A^2}{2} + p_A + \rho g z_A = \rho \frac{V_B^2}{2} + p_B + \rho g z_B \qquad [4.211]$$

$$0 + p_a = \rho \frac{V_{needle}^2}{2} + p_{needle} \qquad [4.212]$$

$$p_{needle} = p_a - \rho \frac{V_{needle}^2}{2} \qquad [4.213]$$

The flow in the needle is undergoing a depression, and this is more the case as the velocity in the needle is increased. For too high velocities, the pressure in the needle becomes lower than the saturating water vapor pressure at the ambient temperature. This causes *cavitation*: a plug of vapor blocks the flow in the needle and therefore stops the aspiration. The filling process is stopped.

As a result, for normal filling, we must have:

$$p_{needle} > p_S \qquad [4.214]$$

And therefore also:

4.1) The equation gives a maximum value of 13,95 m/s for the velocity of fluid in the needle, which results in a value of 8,72 mm/s for the piston.

4.2) The volume flow rate is then equal to 2,74 10-6 m3/s. The time for operation is then 2,9 s

$$V_{needle} < \sqrt{2 \frac{p_a - p_S}{\rho}} \qquad [4.215]$$

Problem 2. Evacuating air:

1) We are dealing with a classic jet problem

We create a frame using an upward vertical axis Oz. We fix the origin O at the level of internal surface of the piston. A jet is a “free” flow, meaning that it does not have any solid borders. For a jet of a perfect fluid, the Bernoulli theorem can be applied. In such a flow, the sum $\rho\frac{V^2}{2}+p+\rho gz=\rho\frac{V^2}{2}+p_G$ is constant. In a free jet flowing into the atmosphere, for reasons of continuity, the pressure is equal to atmospheric pressure. This can be demonstrated rigorously for a vertical jet, which can be treated as a parallel flow (this is an approximation, as the jet is necessarily expanding). In any plane that is perpendicular to the lines of current, so horizontal in this case, p_G is constant, and therefore $p_G=p+\rho gz=Cnst$. On a horizontal plane, z is constant, and therefore, *in the jet*, p is constant. At the interface between the jet and the air, the pressure is continuous. It is therefore equal to atmospheric pressure p_a in the air and in the liquid. This is true regardless of the value of the altitude z :

$$\rho\frac{V^2}{2}+p+\rho gz=\rho\frac{V^2}{2}+p_a+\rho gz \tag{4.216}$$

As a result, V and z vary in opposite directions. There is therefore an altitude for which the jet has a nil velocity. We must look at the limit of this reasoning: in all rigor, the conservation of the flow means that the cross-sectional area increases with the decrease in velocity. The lines of the current are therefore not rigorously parallel. Furthermore, a velocity of zero results in an infinite cross-sectional area. For a calculation of orders of magnitude, which includes any use of the Bernoulli theorem, this reasoning is sufficient. In reality (a water jet in a park, for example), the jet does not keep its integrity; it becomes unstable and breaks into drops. By applying the Bernoulli theorem on a current line between the point A on the surface of the piston on the inside of the body of the syringe and the point B at the top of the jet, we get:

$$\rho\frac{V_A^2}{2}+p_A+\rho gz_A=\rho\frac{V_B^2}{2}+p_B+\rho gz_B \tag{4.217}$$

By inserting the values given for the problem:

$$\rho\frac{V_P^2}{2}+p_P+\rho g(0)=\rho(0)+p_a+\rho g(L+h) \tag{4.218}$$

$$(L+h)=\frac{\rho\frac{V_P^2}{2}+p_P-p_a}{\rho g}=\frac{V_P^2}{2g}+\frac{p_P-p_a}{\rho g} \tag{4.219}$$

$$p_P=p_a+\frac{F}{S}=p_a+\frac{4F}{\pi D^2} \tag{4.220}$$

$$p_P - p_a = \frac{4F}{\pi D^2} = 8276\,Pa \qquad [4.221]$$

$$\frac{p_P - p_a}{\rho g} = 0,844\,m \qquad [4.222]$$

Furthermore, as the force $\vec{F}$ is identical to the one found in question 1, the velocity V at the exit of the needle is identical, and is $V = 4,08\,m.s^{-1}$. The term $\frac{V_P^2}{2g}$ can be deduced, and can be considered negligible:

$$V_P = V\left(\frac{d}{D}\right)^2 = 2,55 10^{-3}\,ms^{-1} \;\; ; \;\; \frac{V_P^2}{2g} = 3,31\,10^{-7}$$

2) As a result, the height of the jet, $L + h$, measured from the internal surface of the piston, is:

$$L + h = \frac{V_P^2}{2g} + \frac{p_P - p_a}{\rho g} \approx \frac{p_P - p_a}{\rho g} = 0,844\,m \qquad [4.223]$$

3) This gives us h as:

$$h = 0,844\,m - L = 0,744\,m = 74,4\,cm$$

EXAMPLE 4.17 (Heron's fountain).–

We are going to analyze and predict the functioning of the device represented on the following page. Invention of the device is credited to Heron (in 120 BC). The dish D and the two reservoirs M and N are filled with water, and consequently a jet appears above D. In this problem, we shall attempt to predict the height of the jet of water. To make things more simple, we shall describe the system and calculate using a simpler geometry of the recipients. The device is made of two cylindrical recipients with diameters of D_R, called M and N. These recipients are plugged and filled partially with water. They are connected to various tubes, and the connections are watertight. A container D, with a diameter of D_R, contains water that has a free surface with the surrounding air, at atmospheric pressure p_a. The parameters a, b and C are the respective initial water levels, measured from the bottom of each recipient, in D, M and N. Three tubes A, B and C go through the system. A links M and N. Both its extremities are in the air. A is only used to equalize the pressures p_N and p_M of the air in both recipients M and N. B links the bottom of D and the liquid contained in N. It is therefore permanently filled with liquid. C has its lower extremity bathed in the liquid contained in M and, at the beginning of the experiment, just reaches the free surface of D. It is also permanently filled with water. $d << D$ is the diameter of C. As we shall demonstrate, the combination of pressures and heights of water will result in a jet of water that comes out of C. We will take as the origin of the altitudes z the horizontal plane associated with the free surface in N *right at the beginning of the experiment*. The free surface in M at the start of the experiment is the applicate h. The free surface of D is the applicate h'. The upper extremity of C is therefore practically at the applicate h'.

1) We consider that the liquids contained in D and N are in equilibrium. Give the relationship between P_a and P_n. Deduce the value of p_n, which *we shall use later in the problem.*

2) By applying the Bernoulli theorem between the free surface of M and the outside, find the applicate Z that the water jet exiting C reaches at the start of the experiment. What is the height h_{Jet} of the jet above the free surface in D? We assume that as it exits the tube C the jet is a perfectly uniform and cylindrical flow. Calculate the flow q_V of this jet.

3) Numerical application. We have:

Density of water $\rho = 1000\, kg.m^{-3}$;

Diameter of D, M and N: $D = 6\, cm$;

Diameter of C: $d = 1\, mm$

$a = 5\, cm$; $b = 4\, cm$; $c = 6\, cm$; $h = 50\, cm$; $h' = 1\, m$.

For simplicity, we shall assume that the value of the flow found at the start of the experiment remains valid during all of the draining of D.

Furthermore, we can consider that inferior extremity of C is practically at the bottom of M. How much time will the jet last?

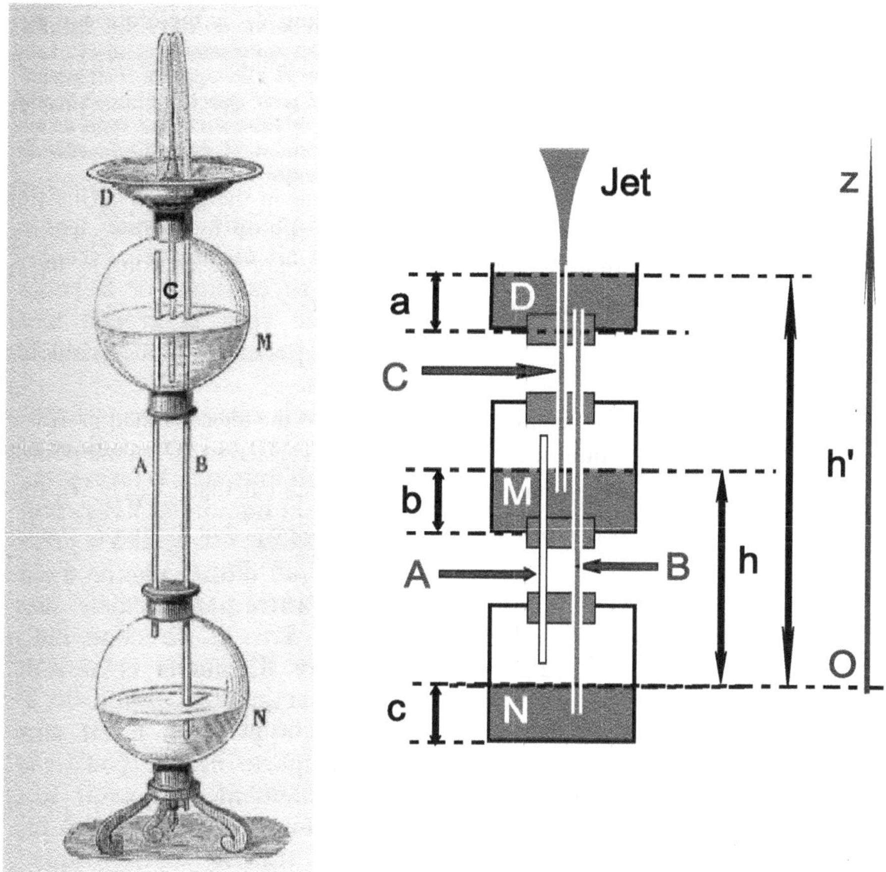

Figure 4.23. *Heron's fountain and its diagram*

Solution:

– Preliminary thoughts:

This device can be puzzling at first. Before attempting any calculations, it is best to try to understand how this "fountain" works. This principle is actually simple. The height of water h' in B creates an overpressure in relation to atmospheric pressure. This pressure is "transmitted" by the tube A to the container M. The free surface of M, which is only at a height of $h - h'$ below D is therefore equipped with potential energy due to the height h. The excess potential energy $\rho g(h' - h)$ can thus be transformed into kinetic energy. If this is understood, the problem becomes a "classic" one.

1) Let us consider the fluid contained in D, N and in the tube B. This fluid forms a continuous volume. Going from the free surface of D to the free surface of N is a drop of h' into a fluid in equilibrium. The pressure passes from p_a to p_N and, according to the fundamental theorem of hydrostatics, there is:

$$p_N = p_a + \rho g h' \qquad [4.224]$$

2) The presence of tube A equalizes the pressures of the air contained in N and in M. By writing the Bernoulli theorem between the surface of M and a point of the jet located above the free surface of D allows us to calculate $V(z)$. We must remind ourselves that the origin of the sides (altitudes) located at the free surface of N:

$$\rho \frac{V_M^2}{2} + p_M + \rho g h = \rho \frac{V^2(z)}{2} + p_a + \rho g z \qquad [4.225]$$

Considering the previous remark and the result obtained in 1:

$$\rho \frac{V_M^2}{2} + p_N + \rho g h = \rho \frac{V^2(z)}{2} + p_a + \rho g z \qquad [4.226]$$

which means:

$$\rho \frac{V_M^2}{2} + p_a + \rho g h' + \rho g h = \rho \frac{V^2(z)}{2} + p_a + \rho g z \qquad [4.227]$$

We note that, considering the conservation of the flow between M and the jet:

$$\frac{V^2(z)}{V_M^2} \approx \frac{D^4}{d^4} << 1 \qquad [4.228]$$

We have:

$$V(z)=2\sqrt{g(h'+h-z)} \quad [4.229]$$

The "theoretical" height h_{jet} of such a jet corresponds to the cote, where the velocity becomes nil:

$$h_{Jet}=h'+h \quad [4.230]$$

We can provide a simple interpretation of this result using energetic terms. The height of water h' has "given" the fluid in N a potential energy $\rho gh'$ per unit of volume. This energy is "transmitted" by the tube A to the recipient M. In the jet, this potential energy is transformed into kinetic energy, which is exchanged with potential energy as the jet goes up. At the summit of the jet, the potential is the same as at the start. The jet therefore "climbs" a height h above the free surface of M, or $h_{Jet}=h'+h$ above the origin of the applicates located at the free surface of N. This provides the flow of the jet, calculated at the exit point of C, or in other words at the applicate $z=h'$:

$$q_V=\pi\frac{d^2}{4}V(h')=\pi\frac{d^2}{2}\sqrt{gh} \quad [4.231]$$

3) At the exit point, the flow of the jet is calculated as:

$$q_V=\pi\frac{d^2}{4}V(h')=\pi\frac{d^2}{2}\sqrt{gh}=\pi\frac{10^{-6}}{2}\sqrt{g*0,5}=3,48.10^{-6}\,m^3.s^{-1} \quad [4.232]$$

The amount of liquid V_{OLM} available in M is:

$$V_{OLM}=\pi\frac{d^2}{4}b=\pi\frac{36.10^{-4}}{4}*4.10^{-2}-1,13.10^{-4}\,m^3 \quad [4.233]$$

This volume is emptied with a flow $q_V = 3,48.10^{-6}\,m^3.s^{-1}$ [4.234]

The duration of the jet is:

$$t_D=\frac{V_{OLM}}{q_V}=\frac{1,13.10^{-4}}{3,48.10^{-6}}=32,5\,s \quad [4.235]$$

When the jet empties the recipient M, the volume of air in this recipient is likely to increase. To maintain the pressure in M, the level in N has to increase again, while the level in D must drop. In all rigor, h' will decrease. At the extreme, its value goes from the initial value $h' = 1m$ to the value $h' - (a + c) = 1 - (0,04 + 0,06) = 0,9m$, which is a relative variation of 10%. It is the effect of this relative variation that is neglected when we assume in question (1) that P_N preserves its value calculated at the start of the experiment.

5

Viscous Fluid Flows: Calculating Head Losses

5.1. Introduction

The hypothesis of an inviscid fluid is useful as it simplifies a great number of problems. This hypothesis involves neglecting the forces that relate to viscosity. These forces are proportional to viscosity gradients in more or less complex ways. This remains true as long as these velocity gradients are low. As such, it is never a fluid that is perfect but rather the flow.

When "the fluid is no longer perfect", the loss of (mechanical) energy due to friction must be accounted for. The energy contained in a flow, as seen before, can be expressed as a head. This concept is useful in quantifying the energy contained in a flow passing through a conduit. The loss of energy along a pipeline is then called a loss of head. This leads to different problems of varying complexities. Two approaches can be taken.

The first, called the "technical" approach and often used in practice, uses empiricism to find a solution. This is the case, for example, when supplying an installation with a fluid through a pipeline network. First, we evaluate the mechanical energy "lost" through friction between the beginning and end of the pipeline, which is expressed as a loss of head.

The second step is to directly calculate this loss of head using a sequence of processes that, except in the very limited case of the application of a laminar flow in a conduit (Poiseuille theory), is based on the use of experimental data. Here, we shall call this a head loss calculation.

A second approach, which is more analytical, considers the structure of the flow and involves the resolution, more or less accurately, of the fluid mechanics equation within a given geometry. This approach is used by researchers or engineers who need to have a better understanding of the structure of a flow when looking at various parameters (velocity profiles, pressure, etc.) that are hidden when using the previous approach. This chapter focuses on the first approach. It looks at the essential elements involved in calculating a loss of head. The following chapter has a more analytical approach and looks at "external" flows, boundary layers, channel flows, etc. It will also consider the non-diagonal terms τ_{ij} of the constant tensor. In a number of problems, a two-dimensional approach is possible, which simplifies the writing of the tangential constraint. This is the case for boundary layers where the orders of magnitude of the gradients are involved. This is also the case in conduits or channels where the gradients become one-dimensional.

5.2. The notion of head: generalized heads

The proof for the Bernoulli theorem showed a total pressure p_T, which was interpreted as the sum of three forms of volume energies. In a flow with no friction, this energy is preserved throughout the flow:

$$p_T = \frac{\rho V^2}{2} + p + \rho g z \qquad [5.1]$$

This energy is expressed as a height through the concept of a head. To achieve this, the total pressure p_T is divided by $\varpi = \rho g$. This results in a head H, which is dimensionally a height:

$$H = \frac{p_T}{\rho g} = \frac{V^2}{2g} + \frac{p}{\rho g} + z = \frac{V^2}{2g} + \frac{p_G}{\rho g} \qquad [5.2]$$

$\varpi = \rho g$ is the density of the fluid. A habit in hydraulics is to pronounce this ω with a bar over it as "pi" (for its resemblance with π). Any other benefits of this practice are not obvious. Let us note at this stage of the definition that the head is applied at a point of the flow. In a perfect fluid, this head is maintained along the whole current line: this is Bernoulli's theorem.

The friction is no longer negligible, mechanical energy is lost along the flow and the head decreases. There is a problem at this point, which is encountered when looking at a flow in a conduit (a laminar flow, for example). This is known as a Poiseuille flow. Let us consider a Poiseuille flow between two sections S_1 and S_2 of a tube. The radius of the tube is *R*, its length is L, the dynamic viscosity of the fluid is μ, its density ρ and its kinematic viscosity is ν. The head loss between two points of a same line of current, one located on S_1 and the other on S_2, is not dependent on the current line considered. It is equal to:

$$\Delta H_{12} = \frac{8\mu Q_V\, L}{\pi R^4\, \rho g} = \frac{8\nu Q_V\, L}{\pi R^4\; g} \qquad [5.3]$$

Determination of this formula requires some knowledge of fluid mechanics, which we shall refresh in due course. However, the head calculated on the same section using this formula will depend on the current considered. It is dependent on *r*, as the p_G is constant on the same section (the flow is parallel but not uniform):

$$p_G = p + \rho g z \qquad [5.4]$$

A head must therefore be defined that is valid over the whole section.

We show that we can therefore define a generalized head.

NOTE.– This generalized head is defined by writing that it must be conserved throughout the flow of a perfect fluid:

$$H_G = \frac{\alpha V_q^{\,2}}{2g} + \frac{p_G}{\rho g} \qquad [5.5]$$

where α is a coefficient that depends on the velocity profile:

$$\alpha = \frac{\iint_S V^3\, dS}{S V_q^3} \qquad [5.6]$$

In practice, in a laminar flow, $\alpha = 2$, and in a turbulent flow (in the case of a "*one – seventh*" profile), $\alpha = 1{,}06$.

It is important to note that this generalized head only makes sense in zones where the flow is locally parallel, which is the case in a tube with an established state, but this is not the case in an establishment area or in most flow singularities. Using this generalized head, we can calculate a loss of head arising between two sections of a hydraulic circuit.

5.3. Practical calculation of a head loss

5.3.1. *Introduction*

The flows found along a circuit are fairly complex. Usually, a circuit is a combination of straight lengths of tube mixed (major head losses) with various singular elements (minor head losses in bends, different changes in the sections, etc.). In straight tubes, the laminar flow of a simple structure is well known in theory and is in a minority compared to turbulent flows, which are also far more complex. Concerning the features of the conduit, where the flow is very disturbed, processing is very intensive, when possible at all. The calculation of head losses is essentially a practical endeavor. The mechanical energy "lost" along the flow must be provided to the circuit in an appropriate manner. Most often, this is achieved using turbomachinery, a ventilator or a pump. In other cases, the energy can be provided by a reservoir (for example, retaining water in the case of a hydroelectric dam).

Note that the term "loss of energy" in no way insinuates an exception to the concept of conservation of energy. The "mechanical" energy is partially transformed into thermal energy as a result of friction. Work is turned into heat and the first principle of thermodynamics is respected. As a result, head losses go beyond "pure" mechanics and enter the realm of thermodynamics. However, in most cases, the temperature increase that results is not noticeable. This is not always the case, for example, when an object enters the atmosphere from space. The heat caused by friction is great enough to create plasma that can lead to the sublimation of the material (for example, of a satellite). This is the reason for the presence of refractory tiles on space ships and the issues caused by them becoming unstuck. Any physicist or engineer, no matter the area, is likely to need to resolve a problem relating to a fluid supply. It is therefore useful to be able to evaluate, even roughly, the head losses of the circuit being supplied so as to choose an adequate pump or ventilator. The precise calculation of a circuit, especially in terms of the so-called "minor" head losses, can require the reading of large hydraulics manuals. Moreover, there are some programs available that are more or less adapted to these situations. We would recommend that professionals stick to these tools; we shall instead give here the occasional user a lighter tool for evaluating head losses, which do not go beyond the limits of this work. The following examples provide information (formulas and data) and a method that we have judged sufficient to size up a circuit, whether in a company or in a laboratory. In a circuit, we can differentiate the linear parts of constant sections (tubes) from the conduit singularities. Note that the loss of head (which is a loss of energy) is often important in singularities. It is therefore vital to pay special attention to these singularities in designing such a circuit. A head loss calculation is consequently divided into two parts

a) The calculation of the “major” loss of head in the straight tube parts, or linear head losses.

b) The calculation of the “minor” loss of head in the conduit singularities, or singular head losses.

5.3.2. *Linear head losses*

We can write the expression of a loss of head (a “generalized” loss, the word is not used in practice, as we shall see later) between two sections. A dimensional analysis provides us with a general form of this loss of head. For a tube, a linear loss of head is defined as J:

$$\Delta H_G = JL \quad [5.7]$$

$$J = \psi \frac{V_q^{\,2}}{2gD} \quad [5.8]$$

where ψ is the linear loss of head coefficient.

To evaluate the loss of head of a linear tube of length L, we must first know the loss of head coefficient ψ, which allows us to calculate the linear loss of head J. To calculate ψ, a distinction is made between smooth tubes and tubes with some degree of roughness (cement tubes, rusty tubes, etc.). For smooth tubes, dimensional analysis shows that the coefficient can be expressed from a single Reynolds number $\psi = \psi(R_D)$. For tubes with a roughness of ε, a dimensionless number $\frac{\varepsilon}{D}$ is defined and ψ becomes a function of two dimensionless numbers: $\psi = \psi\left(R_D, \frac{\varepsilon}{D}\right)$.

ε is an “average value” of roughness, often defined in a standard manner in the hydraulics manuals for a given material. For large values of Reynolds, ψ practically only varies with $\frac{\varepsilon}{D}$.

NOTE.– ε is an "average value" of roughness, often defined in a standard manner in the hydraulics manuals for a given material. There is then $\psi = \psi\left(\frac{\varepsilon}{D}\right)$. This is the case of a "completely rough" tube. In practice, the intermediate solution is called the "partially rough state". Determining $\psi = \psi\left(R_D, \frac{\varepsilon}{D}\right)$ in all these cases can be done using a diagram. Once called the Nikuradse diagram (or harp), today called the Moody diagram, these systems were determined experimentally. In most cases here, we would recommend using formulas that show the data well for both smooth and perfectly rough tubes.

The laminar state is the only case where the formula is the result of a Poiseuille theory ($\alpha = 2$):

$$\psi = \frac{64}{R_D} \qquad [5.9]$$

of the turbulent state.

In a turbulent state, for a smooth tube, Blasius suggests two formulas using the Reynolds numbers ($\alpha = 1{,}06$):

$$R_D < 10^5$$

$$\psi = \frac{0{,}316}{R_D^{0{,}25}} \qquad [5.10]$$

$$R_D > 10^5$$

$$\psi = 0{,}004 + \frac{0{,}25}{R_D^{0{,}25}} \qquad [5.11]$$

In a semi-rough state, Colebrook suggests a formula that takes the diagram into account

$$\frac{1}{\psi} = 1{,}14 - \ 2\log_{10}\left(\frac{\varepsilon}{D} + \frac{9{,}32}{R_D\sqrt{\psi}}\right) \qquad [5.12]$$

In a completely rough state, we can therefore deduce:

$$\frac{1}{\psi}=2\log_{10}\left(\frac{D}{\varepsilon}\right)+1{,}14 \qquad [5.13]$$

In a semi-rough state, we must note that the Colebrook formula is "pessimistic" (it overvalues ψ). Moreover, it has an implicit form, which makes its use uncomfortable. In practice, use of the diagram is recommended.

5.3.3. *Singular loss of head*

For a flow singularity, there is a singular loss of head. To stay in the perspective of a dimensional analysis, it is written as:

$$\Delta H_S=\varsigma\frac{V_q^{\ 2}}{2g} \qquad [5.14]$$

Note that in the Darcy formula, there was $\frac{V_q^{\ 2}}{2gD}$, while here, there is the term $\frac{V_q^{\ 2}}{2g}$. This can be explained by noting that $J=\frac{\Delta H}{L}$ is dimensionless and that ΔH_S is a head with the dimensions of a height. In the case of a section changing singularity, two flow velocities can be calculated. V_q is then the greater of these two flow velocities. In literature, there are more or less extensive lists for the value of ς for different types of singularities. This can result in rather large manuals. Note that the values given in literature mainly involve singularities preceded and followed by considerable lengths of straight tubing. The combination of the two conduit singularities close to one another remains a problem that has not yet been totally resolved. Here, we shall only recall a few classic values for practical evaluation. Various conduit features need to be considered: bends, section changes and orifice plates.

a) For bends, the distinction must be made between various construction methods that influence the value of ξ.

Sharp bend (simple miter-cut joint of two tubes) $\xi=1{,}2$

"Two spot weld" bend (double miter-cut joint at 45°) $\xi=0{,}35$

Rounded bend (more complex construction) $\xi = 0,2$

These values are technically applicable when the bend is preceded by a straight length of tube longer than 40 diameters.

b) Section change: for two section conduit singularity, two flow velocities can be observed. The formulas given use the greater of these two flow velocities (therefore, the one corresponding to the smaller section).

$$\textit{Sudden enlargement } \zeta = \left(1 - \frac{S_1}{S_2}\right)^2 \qquad [5.15]$$

Sudden narrowing: the situation is more complex. The fluid stream contracts from S_1 to $\sigma < S_2$. The loss of head is mostly attributed to the (sudden) section change from σ^2 to S_2. We can use a curve that gives $\frac{S_C}{S_2}$ as a function of $\frac{S_1}{S_2}$ and then the enlargement formula in the form of:

$$\zeta = \left(\frac{S_2}{S_c} - 1\right)^2 \qquad [5.16]$$

Diverging section

In this type of tube, the section is continuously growing. As soon as the angle α between the wall of tube and its axis becomes greater than 7°, a phenomenon of unsticking occurs, like in the case of a sudden change. We can note that $\alpha = 7°$ corresponds to a rather uncommon case. In our simplified approach for the calculation of head losses, we shall limit ourselves to using the sudden enlargement formula, which only slightly overestimates the loss of head observed.

c) Reservoir entrance

A particular case of a sudden section change is the entrance to a reservoir: a tube is linked to a "reservoir" that can be considered to be a tube with a very large (infinite) diameter.

since $S_2 \to \infty$: $\zeta = \left(1 - \frac{S_1}{S_2}\right)^2$ then becomes,

$$\zeta = \left(1 - \frac{S_1}{S_2}\right)^2 \rightarrow 1 \qquad [5.17]$$

The physical interpretation of this result is clear:

$$\Delta H_S = \varsigma \frac{V_q^{\,2}}{2g} = \frac{V_q^{\,2}}{2g} \qquad [5.18]$$

The energy loss is equal to the kinetic part of the head. All the kinetic energy of the fluid is lost at the entrance of the reservoir.

Orifice systems: diaphragms and nozzles

Here again, the construction of these systems is based on standards. Construction based on these standards results in normalized head loss coefficients. In the case of gases, we would advise caution in using this method. Calibration remains in any case a useful precaution. For a diaphragm with an orifice of s centered on a tube of section S, we assume that the fluid stream contracts to an area of $\sigma < s$. Next, we determine the head loss coefficient in the same way as for a sudden enlargement between σ and S, which is

$$\zeta = \left(\frac{\sigma}{S} - 1\right)^2 \qquad [5.19]$$

In terms of our simplified approach, we determine σ using:

$$\frac{\sigma}{S} = 0{,}63 + 0{,}37\left(\frac{s}{S}\right)^3 \qquad [5.20]$$

5.4. Circuit calculations

Here, we provide a limited number of examples that show the main aspects of head loss calculations. In Examples 5.1–5.3, we shall practice doing head loss calculations between the two ends of a circuit. Examples 5.4 and 5.6 involve head losses in a supplied system, where the energy is provided by a reservoir (Examples 5.4 and 5.5) or, and this is more original, forces of inertia (Example 5.6).

EXAMPLE 5.1.–

The circuit AE is represented below and is made up of smooth circular conduits.

The main characteristics of this conduit are as follows:

Horizontal conduit AB: diameter $D_A = 1\,cm$ and length $AB = 50\,m$.

Vertical conduit BC: diameter $D_B = 1\,cm$ and length $BC = 15\,m$.

Horizontal conduit CD: diameter $D_C = 1\,cm$ and length $CD = 50\,m$.

Horizontal conduit DE: diameter $D_E = 2\,cm$ and big length (the tube goes far beyond E and beyond the drawing). We have $DE = 100\,m$.

At B and C, the bends are rounded.

At D, the section change is sudden.

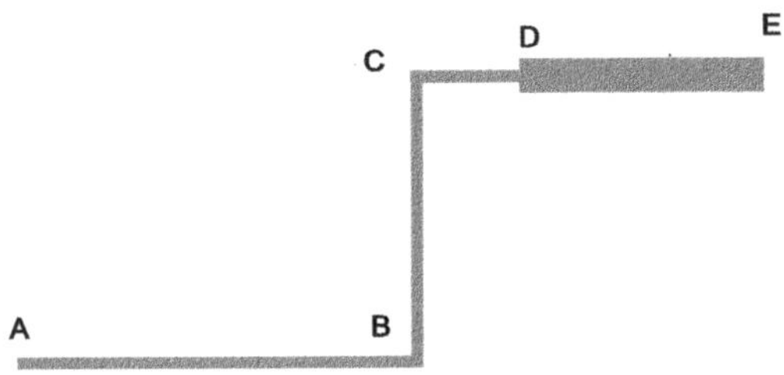

Figure 5.1. *Hydraulic circuit from Example 5.1*

The circuit is crossed by a flow of water of $q_v = 0{,}1571\ litres.s^{-1}$.

The static pressure in A is $p_A = 60\,bar$.

The origin of the applicate z is at the level of point A.

Information for the water: dynamic viscosity is $\mu = 1{,}8.10^{-3}\,Pl$ and density is $\rho = 1000\,kg.m^{-3}$.

A) First, we assume that the fluid is perfect.

What is the static pressure in B, C and E?

B) Second, we recognize the fact that the fluid is real.

1) What are the flow velocities in the sections AD and DE?

2) In the part AD, is the flow laminar or turbulent?

3) In part DE, is the flow laminar or turbulent?

4) Calculate the following:

– the linear head losses in the different stretches;

– the singular head losses;

– the total head loss between points A and E.

5) What is the head at A (express it in m)? Deduce the head at E.

6) What is the static pressure at E? Compare with the result obtained for a perfect fluid.

Solution:

A) The hypotheses required for the Bernoulli theorem are verified.

The energy is therefore conserved per unit of volume between the points A, B, C, D and E:

$$\rho\frac{V_A^2}{2} + p_A + \rho g z_A = \rho\frac{V_B^2}{2} + p_B + \rho g z_B = \rho\frac{V_C^2}{2} + p_C + \rho g z_C$$

$$= \rho\frac{V_D^2}{2} + p_D + \rho g z_D = \rho\frac{V_E^2}{2} + p_E + \rho g z_E \quad [5.21]$$

which can furthermore be written in a perfectly equivalent form in terms of the conservation of the head:

$$\frac{V_A^2}{2g} + \frac{p_A}{\varpi} + z_A = \frac{V_B^2}{2g} + \frac{p_B}{\varpi} + z_B = \frac{V_C^2}{2g} + \frac{p_C}{\varpi} + z_C \quad [5.22]$$

$$= \frac{V_D^2}{2g} + \frac{p_D}{\varpi} + z_D = \frac{V_E^2}{2\rho g} + \frac{p_E}{\varpi} + z_E;\ \varpi = \rho g \quad [5.23]$$

Moreover, there is

$$S_A = \pi \frac{D_A^2}{4} = 7{,}85.10^{-5}\ m^2;\ V_A = \frac{q_V}{S_A} = \frac{0{,}1571.10^{-3}}{7{,}85.10^{-5}} = 2\,m.s^{-1};\ V_A = V_B = V_C$$

$$S_D = \pi \frac{D_D^2}{4} = \pi \frac{4.10^{-4}}{4} = 3{,}14.10^{-4}\ m^2$$

$$V_D = \frac{q_V}{S_A} = \frac{0{,}1571.10^{-3}}{3{,}14.10^{-4}} = 0{,}5 m.s^{-1} = 50 cm.s^{-1}$$

$$V_D = V_E;\ z_A = z_B = 0;\ z_C = z_D = z_E = 15m$$

Resulting in:

$$\rho\frac{4}{2} + p_A + 0 = \rho\frac{4}{2} + p_B + 0 = \rho\frac{4}{2} + p_C + \rho g * 15 \qquad [5.24]$$

$$= \rho\frac{0{,}25}{2} + p_D + \rho g * 15 = \rho\frac{0{,}25}{2} + p_E + \rho g * 15$$

Finally, we find:

$$p_B = p_A = 60 bar = 6.10^6\ Pa \qquad [5.25]$$

$$p_E = \rho\frac{4}{2} + p_C - \rho\frac{0{,}25}{2} = 58{,}53.10^5 + 1000 * \frac{3{,}75}{2} = 58{,}55.10^5\ Pa = 58{,}55 bar$$

B) The fluid is real.

1) The calculation of flow velocities is the same as in a perfect fluid:

$$V_A = V_B = V_C = 2\,m.s^{-1}$$

$$V_D = V_E = 50 cm.s^{-1}$$

2) To determine the state of the flow, we calculate the kinematic viscosity ν, followed by the Reynolds number R_{DA}:

$$\nu = \frac{\mu}{\rho} = \frac{1,8.10^{-3}}{1000} = 1,8.10^{-6}\ m^2.s^{-1}$$

$$R_{DA} = \frac{V_A D_A}{\nu} = \frac{2*10^{-2}}{1,8.10^{-6}} = 1,11.10^4$$

The Reynolds number is far greater than 2000. The flow is turbulent.

3) Similarly, in the section DE, we calculated the Reynolds number R_{DE}:

$$R_{DE} = \frac{V_E D_E}{\nu} = \frac{0,5*2.10^{-2}}{1,8.10^{-6}} = 5556$$

The Reynolds number is far greater than 2000. The flow is turbulent.

4) Evaluation of the loss of head

a) The linear loss of head of the different stretches

The tube is smooth. In a turbulent state, we can use the Blasius formulas to determine the linear head loss coefficients ψ and the linear head losses:

$$R_D < 10^5$$

$$\psi = \frac{0,316}{R_D^{0,25}}$$

$$J = \psi \frac{V_q^2}{2gD}$$

On the stretch AD:

$$\psi = \frac{0,316}{R_D^{0,25}} = \frac{0,316}{\left(1,11.10^4\right)^{0,25}} = 3,08.10^{-2}$$

$$J = \psi \frac{V_q^2}{2gD} = 3,08.10^{-2} \frac{4^2}{2g*10^{-2}} = 2,51$$

On the stretch DE :

$$\psi = \frac{0,316}{R_D^{0,25}} = \frac{0,316}{(5556)^{0,25}} = 3,66.10^{-2}$$

$$J = \psi \frac{V_q^2}{2gD} = 3,66.10^{-2} \frac{0,5^2}{2g*2.10^{-2}} = 2,33$$

The total loss of head is therefore:

$$\Delta H_{Lin} = 2,51*(AB + BC + CD) + 2,33*DE = 2,51*115 + 2,33*100$$

b) Singular head losses

Singular losses of head link two bends and a sudden enlargement.

The two bends have the same diameter through which a flow travels that has a velocity of

$$V_A = V_B = V_C = 2\,m.s^{-1}$$

The head loss coefficient is $\zeta = 0,2$ for a rounded bend. The loss of head for each bend is therefore:

$$\Delta H_{SC} = \zeta \frac{V_A^2}{2g} = 0,2 \frac{4}{2g} = 4,08.10^{-2}\,mCg$$, which is low.

The head loss coefficient for the sudden enlargement is calculated as:

$$\zeta = \left(1 - \frac{S_A}{S_E}\right)^2 ; \quad \zeta = \left[1 - \left(\frac{D_A}{D_E}\right)^2\right]^2 = (1 - 0,25)^2 = 0,562$$

The loss of head by sudden enlargement is therefore:

$$\Delta H_{SEL} = \zeta \frac{V_A^2}{2g} = 0,562 \frac{4}{2g} = 0,115\,mCg$$

c) The total loss of head between the points A and E

The total loss of head between A and E is therefore:

$$\Delta H = \Delta H_{Lin} + 2 * \Delta H_{SC} + \Delta H_{SEL} = 521{,}6 + 0{,}04 + 0{,}115 = 521{,}76\, mCE$$

5) By definition of the head and considering the origin of the altitudes at the level of A:

$$H_A = \frac{V_A^2}{2g} + \frac{p_A}{\varpi} + z_A;\ \varpi = \rho g = 9810$$

$$H_A = \frac{4}{2g} + \frac{60.10^5}{9810} + 0 = 611{,}82\, mCE$$

Note that the head maximum is contained in the potential energy of the pressure. Moreover, the singular head losses are low in terms of the linear head losses. The flow velocities are low for the length of the tubes. The head at E is such that

$$H_E = 611{,}82 - 521{,}76 = 90{,}1\, mCE$$

$$H_A = H_E + \Delta H \qquad [5.26]$$

6) By the definition of the head,

$$H_E = \frac{V_E^2}{2g} + \frac{p_A}{\varpi} + z_E = \frac{(0{,}5)^2}{2g} + \frac{p_E}{9810} + 15 = 90{,}1m$$

as $z_E = BC = 15\, m$.

The pressure p_E at E is therefore $p_E = 9810 * \left(90{,}1 - 15 - 1{,}27.10^{-2}\right)$

$p_E = 7{,}36.10^5\ Pa = 7{,}36\, bar$. This result can be compared with the pressure found under the hypothesis of a perfect fluid, which is $p_E = 5{,}855.10^6\ Pa = 58{,}55\, bar$.

The difference: $\Delta\, p_E = 58{,}55 - 7{,}498 = 51{,}05\, bar = 511{,}9.10^5\ Pa$

The pressure difference is due to the loss of mechanical energy in a real fluid.

Translated as a head, there is $\Delta p_E = 51{,}05\,bar$, which corresponds to $\Delta H = \frac{\Delta p_E}{\varpi} = \frac{51{,}19.10^5}{9810} = 521{,}8m$, which does indeed – with the adequate rounding – correspond to the loss of energy in the pipeline.

EXAMPLE 5.2 (Thermal supply circuit).–

A circuit designed to be a heat exchange installation is made up of a smooth tube of internal diameter $16\,mm$, with the geometry shown in the figure. The bends are rounded. A flow of water goes through it at a rate of $q_V = 12\,liters.mn^{-1}$. Note that the density of water is $\rho = 1000\,kg.m^{-3}$ and its viscosity is $\mu = 1\,cps$.

1) What is the loss of head ΔH_{T1} in this circuit? What is the pressure difference Δp_1 between the entrance and the exit?

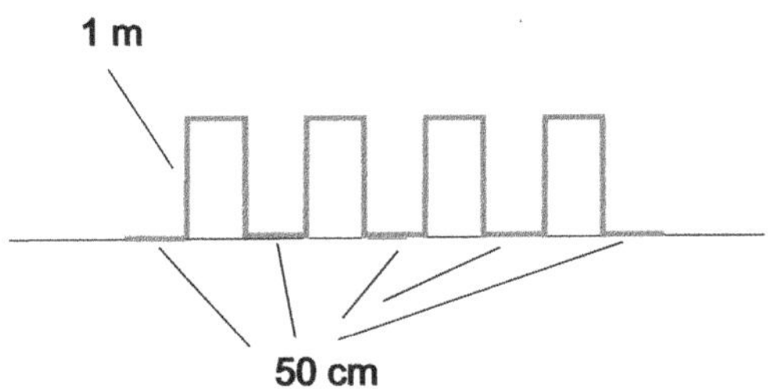

Figure 5.2. *Thermal supply circuit*

2) This circuit is linked to a straight pipeline of internal diameter 32 mm. What is the total loss of head for the whole system for a flow of q_V? We shall neglect the effects of the length of the 32-mm-diameter section.

Solution:

1) The first step in any calculation of the loss of head involves the determination of the flow velocity V_q and of the Reynolds number calculated in all of the straight parts of the circuit:

$$q_V = sV_q = \frac{12.10^{-3}}{60} = 2.10^{-4}\,m^3 s^{-1};\; s = \pi\frac{d^2}{4} = 2{,}01.10^{-4}\,m^2$$

$V_q = 0{,}995\, ms^{-1}$. The Reynolds number requires knowledge of the kinematic viscosity of the fluid, expressed in SI, which is m^2s^{-1}. μ in centipoise shows a dynamic viscosity expressed in a cgs system. Let us remind ourselves of the conversion into SI:

$$1\,Poiseuille = 10\ poises = 1000\,centipoises$$
$$1\,Pl = 10\ ps = 1000\ cps$$

ν is the kinematic viscosity of the fluid and R_D is the Reynolds number, which has the same value in all the straight section of the circuit since section s is constant:

$$\nu = \frac{\mu}{\rho} = \frac{10^{-3}}{10^{3}} = 10^{-6}\ m^2 s^{-1}$$

$$R_D = \frac{V_q\, d}{\nu} = \frac{0{,}995.16.10^{-3}}{10^{-6}} = 1{,}592.10^{4}$$

The flow is in a turbulent state.

The circuit is made up of:

– eight linear stretches of 1 m;

– five linear stretches of 0.5 m;

– 16 rounded bends.

The loss of head is therefore made up of:

– a linear loss of head for a length of piping equal to $L = 10{,}5\,m$;

– a singular loss of head as a result of 16 bends.

The linear loss of head ΔH_L is calculated using the Darcy formula:

$$\Delta H_L = JL;\ J = \psi \frac{V^2}{2gD} \qquad [5.27]$$

ψ , the linear head loss coefficient, is calculated using the Blasius formula, valid for $R_D < 10^5$:

$$\psi = \frac{0,316}{R_D^{0,25}} = 2,81.10^{-2}$$

$$\Delta H_L = \psi \frac{V_q^2}{2gd} L = 2,81.10^{-2} \frac{(0,995)^2}{2g16.10^{-3}} 10,5 = \frac{0,316.\nu^{0,25}}{2g\,d^{1,25}} V_2^{1,75} L \qquad [5.28]$$

$\Delta H = 0,93\,mwg$. Here, we have used the classic notation for mwg , and we must remember that the head is expressed in meters of water gauge column. The singular head losses ΔH_s are expressed for each bend using a singular head loss coefficient ξ :

$$\Delta H_s = \xi \frac{V_q^2}{2g} \qquad [5.29]$$

Note that the d present in the denominator in expression J has disappeared here. Indeed, the linear loss of head is defined as a unit of length. It is therefore dimensionless. A singular loss of head, like any head, has the dimension of a length. The head loss coefficient ξ is here taken as being equal to 0.2. Each bend is preceded by a length equal to more than 60 diameters. For the 16 bends, the total singular loss of head is:

$$\Delta H_s = 16\xi \frac{V_q^2}{2g} = 16.0,2\frac{(0,995)^2}{2g} = 0,16\,mwg$$

The total loss of head is therefore:

$$\Delta H_{T1} = \Delta H_L + \Delta H_S = 0,93 + 0,16 = 1,09\,mwg \qquad [5.30]$$

By the definition of heads, and with e and s being indices attached to the entrance and exit of the circuit.

The entrance and the exit have the same cote, and the fluid there has the same velocity, so therefore the difference in pressure is:

$$\Delta H_{T1} = H_e - H_s = \frac{V_e^2}{2g} + \frac{p_e}{\varpi} + z_e - \frac{V_s^2}{2g} + \frac{p_s}{\varpi} + z_s \quad [5.31]$$

$$z_e = z_s \ ; \ V_e = V_s = V_q \ ; \ \frac{\Delta p_1}{\varpi} = \frac{p_e}{\varpi} - \frac{p_s}{\varpi} = \Delta H_{T1} \quad [5.32]$$

$$\Delta p_1 = \rho g \, \Delta H_{T1} = 1{,}06.10^4 \, Pa = 0{,}106 \, bar \quad [5.33]$$

2) If the linear loss of head in the tube is neglected, the only extra loss of head is a singular head loss caused by a sudden section change of s to S :

$$s = \pi \frac{d^2}{4} = 2{,}01.10^{-4} \, m^2 \qquad S = \pi \frac{D^2}{4} = 8{,}04.10^{-4} \, m^2$$

The linear head loss coefficient is $\zeta = \left(1 - \frac{s}{S}\right)^2 = 0{,}562$.

And the extra loss of head is $\zeta = \left(1 - \frac{s}{S}\right)^2 = 0{,}562$.

$\Delta H_s = \xi \frac{V_q^2}{2g} = 2{,}836.10^{-2} \, mCE$. As a result, the new total loss of head and the new pressure difference between the upstream and the downstream arc

$$\Delta H_{T2} = 1{,}09 + 2{,}836.10^{-2} = 1{,}11 \, mCE$$

Δp_2 is deduced easily as $\Delta p_2 = \rho g \, \Delta H_{T2} = 1{,}09.10^4 \, Pa = 0{,}109 \, bar$

EXAMPLE 5.3 (Calculation in a wind tunnel).–

A wind tunnel has the shape of a tube AB with a diameter $D_1 = 20 \, cm$ and a length $L_{AB} = 10 \, m$. This tube is supplied in air by a centrifugal ventilator through a circuit that is located on the same horizontal plane as the tunnel.

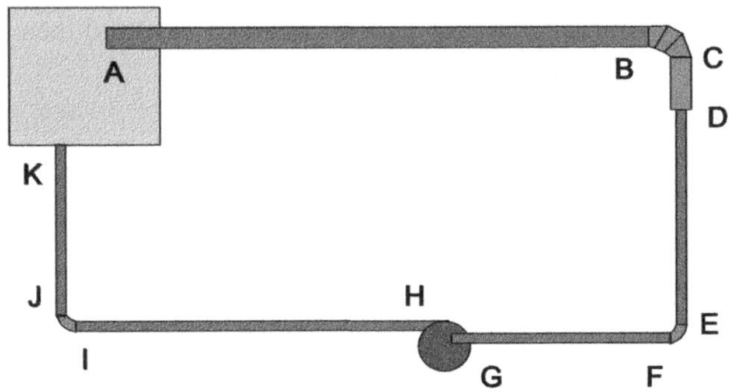

Figure 5.3. *Diagram of the circuit*

Upstream, the tunnel is supplied by a settling chamber: The fluid therefore goes into a reservoir R that is sufficiently large for the velocity within it to be practically equal to zero.

This chamber is supplied by a pipeline with a diameter of $D_2 = 10\,cm$ made up of two straight elements of respective lengths $L_{JK} = 3\,m$ and $L_{HI} = 7\,m$, separated by a bend linking the chamber and the exit of the ventilator. The tube is linked to the chamber by a simple flange.

The extremity B of the tunnel is linked by a bend to a straight section of diameter $D_1 = 20\,cm$ and of length $L_{CD} = 1\,m$. A sudden section change links this pipeline of diameter $D_1 = 20\,cm$ to another pipeline of diameter $D_2 = 10\,cm$ and with a length of $L_{DE} = 4\,m$ and then to a bend of the same diameter. A length of pipeline of $L_{FG} = 4\,m$ with a diameter $D_2 = 10\,cm$ links the exit of this bend to the entrance of the ventilator.

To be frugal, the bends only have two welds.

The tunnel is supplied with an airflow of $Q_V = 6000\,m^3.hr^{-1}$.

1) What is the difference in head between the extremities G and H of the circuit?

2) Between its flanges A and B, the ventilator provides a pressure equal to $p_{HG} = 15000\,Pa$ for the flow considered. The entrance and exit velocities of the ventilator are considered to be the same.

Can the circuit be supplied properly?

Otherwise, what modifications could be suggested for the circuit?

Solution:

1) Calculation of the head loss ΔH_{AB}.

In any calculation of a head loss for a given flow, some preliminary calculations are essential

Calculation of the sections:

Tube of diameter $D_1 = 20cm$, $S_1 = \pi \frac{(0,2)^2}{4} = 3,14.10^{-2} m^2$.

Tube of diameter $D_2 = 10cm$, $S_2 = \pi \frac{(0,1)^2}{4} = 7,85.10^{-3} m^2$

Calculation of the flow velocities:

$$Q_V = 6000\, m^3.hr^{-1} = 1,67\, m^3.s^{-1} \quad [5.88]$$

Tube of diameter $D_1 = 20cm$, $V_{q1} = \frac{1,67}{3,14.10^{-2}} = 53,1 m.s^{-1}$

Tube of diameter $D_2 = 10cm$, $V_{q1} = \frac{1,67}{7,85.10^{-3}} = 212,7 m.s^{-1}$

Calculation of the Reynolds numbers:

Kinematic viscosity $\nu = \frac{1,8.10^{-5}}{1,3} = 1,38\, m^2.s^{-1}$

Tube of diameter $D_1 = 20cm$, $R_{D1} = \frac{V_{q1} D_1}{\nu} = \frac{53,1 * 0,2}{1,38.10^{-5}} = 7,69.10^5$

The flow is turbulent.

Tube of diameter $D_2 = 10cm$, $R_{D2} = \frac{V_{q2} D_2}{\nu} = \frac{212,7 * 0,1}{1,38.10^{-5}} = 1,54.10^6$

The flow is turbulent.

a) Straight pipelines

Calculation of the linear head loss coefficients using the Darcy formula:

$J = \frac{V_q^2}{2gD}$. Since the flow is turbulent throughout and the tubes are smooth, we can use the Blasius formula:

$$R_D < 10^5 \; ; \; \psi = \frac{0{,}316}{R_D^{0{,}25}} \; ; \; R_D > 10^5 \; ; \; \psi = 0{,}004 + \frac{0{,}25}{R_D^{0{,}25}}$$

Tube of diameter $D_1 = 20\,cm$

$$\psi_1 = 0{,}004 + \frac{0{,}25}{R_{D1}^{0{,}25}} = 1{,}24.10^{-2}$$

$$J_1 = \psi_1 \frac{V_{q1}^2}{2g\,D_1} = 8{,}91$$

Tube of diameter $D_2 = 10\,cm$

$$\psi_2 = 0{,}004 + \frac{0{,}25}{R_{D2}^{0{,}25}} = 1{,}11.10^{-2}$$

$$J_2 = \psi_2 \frac{V_{q2}^2}{2g\,D_2} = 255{,}9$$

In total, the linear head losses are therefore as follows:

Tube of diameter $D_1 = 20\,cm$, $L_1 = 10 + 1 = 11\,m$

$$\Delta H_{Lin1} = J_1\,L_1 = 11 * 8{,}91 = 98\,m\,Cg$$

Cg here means gas column (when it is a liquid, it is written as CE)

Tube of diameter $D_2 = 10\,cm$, $L = 4 + 4 + 7 + 3 = 18\,m$

$$\Delta H_{Lin2} = J_2\, L_2 = 18 * 255{,}9 = 4606\,Cg$$

b) Singular head losses

The bends have two welds; therefore, $\zeta = 0{,}3$.

A bend with a diameter of $D_1 = 20\,cm$, $V_{q1} = 53{,}1\,m.s^{-1}$

$$\Delta H_{SC1} = \varsigma \frac{V_{q1}^{\;2}}{2g} = 43{,}11\,mCg$$

Tube of diameter $D_2 = 10\,cm$, $V_{q1} = 212{,}7\,m.s^{-1}$

$$\Delta H_{SC2} = \varsigma \frac{V_{q2}^{\;2}}{2g} = 691{,}76\,mCg$$

Section changes

Reservoir entrance: $\Delta H_{\text{Res}} = 1 * \frac{V_{q2}^{\;2}}{2g} = 2306\,mCg$

Section change at D: $\Delta H_{SCS} = \zeta \frac{V_{q2}^2}{2g} = 1{,}5 \frac{212{,}7^2}{2g} = 3459\,mCg$

Tunnel entrance: we can neglect the head loss caused by the entering tube.

The total loss of head between G and H is therefore:

$$\Delta H_{GH} = \Delta H_{Lin1} + \Delta H_{Lin2} + \Delta H_{SC1} + 3 * \Delta H_{SC2} + \Delta H_{\text{Res}} + \Delta H_{SCS}$$

$$\Delta H_{GH} = 98 + 4606 + 43{,}11 + 3 * 691{,}8 + 2306 + 3459 = 12588\,mCg$$

The entrance and exit velocities of the fan are the same. There is no change in altitude in the problem, so the difference in head between the entrance and exit of the fan is equal to the difference in pressure. Indeed:

$$\Delta H_{HG} = \frac{p_H - p_G}{\rho g} + \frac{V_{qH}^2 - V_{qG}^2}{2g} + (z_H - z_G) = \frac{p_H - p_G}{\rho g} + \frac{0}{2g} + (0)$$

$$\Delta p_{HG} = p_H - p_G = \rho g \Delta H_{HG}$$

ΔH_{GH} gives us an expected pressure difference between H and G of:

$$\Delta p_{HG} = \rho g \Delta H_{HG} = 1{,}3*9{,}81*12588 = 1{,}605.10^5 \ Pa$$

This difference is far greater than the pressure $p_{HG} = 15000 \ Pa$ *available to the fan*. In terms of head, the ventilator provides a head of

$$\Delta H_{HG} = \frac{15000}{\rho g} = 1176 \, mCg$$

Several solutions can be considered:

1) *Replace the fan with a blower or a compressor.* This is the most costly solution and not the most rational.

2) *Look for possible excessive energy consumption, and address this.*

Note that a single tunnel only "costs" $98m$ "*gazgauge*".

To decrease the loss of head, we could envisage constructing the return to the ventilator through a diameter with the same diameter of the tunnel, that is, with $L = 29m$ and with a diameter of $D_1 = 20cm$. The flow velocity is then $V_{q1} = 53{,}1 m.s^{-1}$ in all of the circuit, the three bends have the same loss of head ΔH_{SC1} $43{,}11m$ and $\Delta H_{SCS} = 3459m$ disappears. Only the loss at the entrance of the reservoir ΔH_{Res} remains. However, its value is lower since the flow velocity is divided by 4. There is $\Delta H_{Res} = 1*\frac{V_{q1}^2}{2g} = 143{,}7\,m$.

The total loss of head becomes:

$$\Delta H_{GH} = \Delta H_{Lin1} * \frac{29}{10} + 3 * \Delta H_{SC1} + \Delta H_{Res} \quad [5.34]$$

$$\Delta H_{GH} = 98 * \frac{29}{10} + 3 * 43{,}11 + 143{,}7 = 284{,}2 + 129{,}33 + 143{,}7 = 557{,}2$$

This value is compatible with the energy available for the fan, $\Delta H_{HG} = 1176 m$. An additional head must be added to regulate the velocity, which will also allow us to better regulate the study tunnel for later studies. Note that, practically speaking, the entrance and the exit of the ventilator must be adapted to the new section of tube. We can think of a very long diverging length and a very long converging length, which is permitted within the dimensions of the system. Note that the respective lengths $L_{FG} = 4m$ and $L_{HI} = 7m$ correspond to the respective converging angles of $\alpha = 0{,}7°$ and $L_{FG} = 0{,}409°$. This is far below 7°, and the loss of head will be nearly identical to the loss within the tube. Two converging lengths of 4 m are sufficient with a joint welded on to the 7 m length. From a monetary point of view, the converging length must still be constructed. However, the extra cost of an 18 m tube with a diameter of 20 cm and of converging lengths is far more reasonable than the cost associated with buying (or simply using) a compressor.

EXAMPLE 5.4 (Diagram of a hydroelectric dam).–

A reservoir with a depth of $h = 20m$ supplies a pipeline $H = 100m$ over a drop BC. The pipeline BC makes an angle $\alpha = 45°$ with the horizontal. The pipeline is smooth, with an internal diameter of $D = 20cm$. The density of water is $\rho = 1000 kg.m^{-3}$ and its viscosity is $\nu = 10^{-6} m.s^{-2}$. Atmospheric pressure is $p_a = 1bar$.

First of all, we assume that the fluid is perfect.

1) What would then be the flow velocity V_{q1} and the volume flow q_{V1}?

The fluid is now real.

2) First of all, a hypothesis must be made regarding the state of the flow. For this, we use the flow velocity found in 1.

3) Remember the definition of the head in one point. We take the origin of the cote at C. Give the value of the head at point D located at the free surface of the reservoir.

4) Write the expression of H_C at C as a function of the flow velocity V_q (still unknown), knowing that the tube opens to C at atmospheric pressure.

5) Find the expression for the loss of head $\Delta H_{BC}(V_q)$ in the tube BC as a function of the flow velocity V_q, which is still unknown.

We shall neglect the head losses in the zone AB of the tube as well as the singular head losses caused by the section change at A and by the direction change at B.

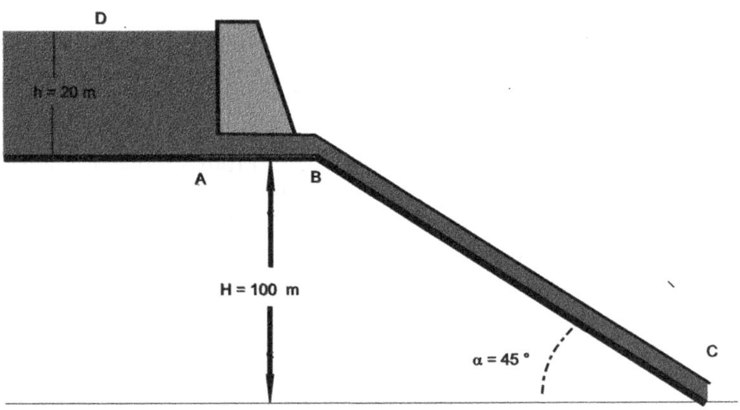

Figure 5.4. *The hydraulic dam*

6) Using the expression for the loss of *head* $\Delta H(V_q)$ written in (5), find the observed value of the flow velocity V_{q2} and that of the volume flow q_{V2} going through tube AC.

NOTE.– *We need to resolve an algebraic equation containing two different powers of V_q. To simplify the calculations, we can use an approximate resolution, using $V_q^2 \approx V_q^{1,75}$.*

7) Verify post-resolution that the hypothesis made regarding the state of the flow is indeed correct.

Solution:

1) The fluid is perfect. The calculation is classic. This goes back to the result of the Torricelli formula:

$$V=\sqrt{2g(H+h)}=48{,}51\,m.s^{-1} \qquad q_V=SV\ ;\ S=\pi\frac{D^2}{4}=3{,}14.10^{-2}\,m^2$$

$$q_{V1}=1{,}52\,m^3.s^{-1}$$

The fluid is now real.

2) By using $V=48{,}51\,m.s^{-1}$, we can calculate the Reynolds number. ν is expressed in $m^2.s^{-1}$ and is therefore a kinematic viscosity:

$$V=48{,}51\,m.s^{-1} \qquad R_D=\frac{VD}{\nu}=9{,}7.10^6$$

For this flow generated from a constant energy reservoir, the velocity observed in the presence of head losses will be necessarily lower than $V=48{,}51\,m.s^{-1}$ due to these same head losses. Considering the Reynolds number, which here is greatly superior to 2000, we can suggest the hypothesis that the flow is turbulent without too great a risk.

NOTE.– Once again, we can see why it is necessary – as stated above – to know the units, both SI and CGS, for the two viscosities.

3) By the definition of the *head*, at a point of the surface of the reservoir, located at a cote $z_D=120\,m$, immobile and under atmospheric pressure, there is:

$$H_D=\frac{V^2}{2g}+\frac{p_D}{\varpi}+z_D=0+\frac{10^5}{\varpi}+120 \qquad H_D=130{,}19\,mCE$$

4) By the definition of the *head*, at C, we can write:

$$H_C=\frac{V_q^2}{2g}+\frac{p_C}{\varpi}+z_C=\frac{V_q^2}{2g}+\frac{p_a}{\varpi} \qquad [5.35]$$

$$H_C=\frac{V_q^2}{2g}+10{,}19 \qquad [5.36]$$

5) We only consider linear *head* losses

$$J=\psi\frac{V_q^2}{2gD} \tag{5.37}$$

for this smooth tube $\psi=\frac{0,316}{R_D^{0,25}}$. The length of the tube is $L=100*\cos\frac{\pi}{4}=141,4\,m.$

The loss of *head* $\Delta H\left(V_q\right)$, expressed as a function of V_q (still unknown), is:

$$\psi=\frac{0,316}{\left(\dfrac{V_q\,0,2}{10^{-6}}\right)^{0,25}}=\frac{1,494.10^{-2}}{\left(V_q\right)^{0,25}}\ ;\ J=\frac{1,494.10^{-2}}{\left(V_q\right)^{0,25}}\frac{V_q^2}{2g0,2}=3,807.10^{-3}\,V_q^{1,75}$$

$$\Delta H=J\,L=3,807.10^{-3}*141,4*V_q^{1,75}=0,538*V_q^{1,75}$$

6) The loss of *head* decreases the head between D and C:

$$H_D=H_C+\Delta H$$

By combining the expressions established above:

$$130,9=\frac{V_q^2}{2g}+10,19+0,538*V_q^{1,75} \tag{5.38}$$

So the equation to solve is:

$$5,110^{-2}V_q^2+\ 0,538*V_q^{1,75}=120,71$$

By accepting the approximation of $V_q^2\approx V_q^{1,75}$,

seeing that the coefficient of V_q^2 is lower than that of $V_q^{1,75}$, we shall use the approximation in the form of $0,589*V_q^{1,75}\approx 120,71\ \ V_q\approx 20,94\,m.s^{-1}$

A numerical resolution of the equation would give us $V_q=19,87\,m.s^{-1}$.

A simple solver on a calculator is enough for this calculation. However, we have deemed it worthwhile to recommend the use of an approximation that is quite simple, following proper use of the coefficients of V_q^2 and $V_q^{1,75}$.

The volume flow becomes $q_{V2} = SV = 3{,}14.10^{-2} * 20{,}94 = 0{,}657\, m^3.s^{-1}$.

7) The Reynolds number is then

$R_D = \dfrac{20{,}94 * 0{,}2}{10^{-6}} = 4188$. The flow is indeed turbulent.

EXAMPLE 5.5 (Supplying a circuit through gravity).–

A tube with a diameter of $d = 4\,cm$ and a length of $L = 20\,m$ is linked to a reservoir filled with water and with a very large diameter. The height of the water in the reservoir above the orifice of the tube is $h = 5\,m$. In this problem, this height is assumed to be invariable. The density of water is $\rho = 1000\, kg.m^{-3}$ and its viscosity is $\mu = 1\, centipoise$.

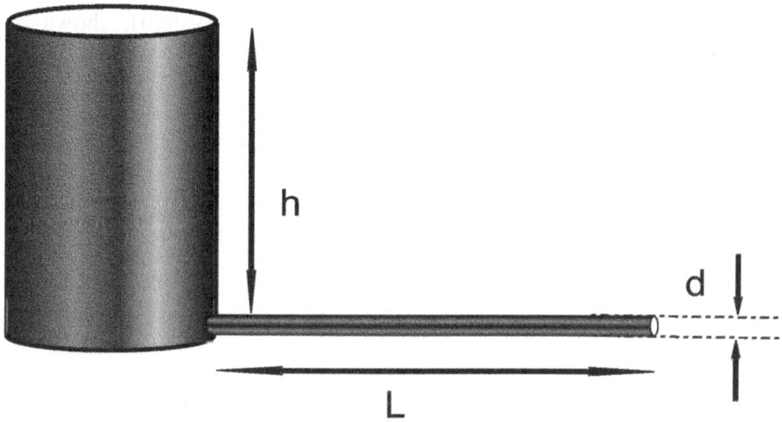

Figure 5.5. *Supplying a circuit with a reservoir*

1) We hypothesize that the fluid is perfect. Calculate the flow q_{V1} of water in the tube, assuming this hypothesis.

2) We now consider the viscosity of the fluid. Using the result from 1, determine whether the flow state actually observed is laminar or turbulent.

3) Write the expression of the head at the free surface of the reservoir and at the exit point of the tube. Write the expression of the head difference between these two points as a function of the flow velocity V_q in the tube. Deduce from this the value of V_q and the real draining flow rate through the tube.

Solution:

1) This is in the context of the Bernoulli theorem hypotheses. The drainage problem is a classic one. We can refer here to the chapter dedicated to perfect fluids flows. We also see the Torricelli problem again here. Taking into account the hypotheses, the exit velocity is $V_1 = \sqrt{2gh}$.

s is the horizontal section of the tube. Calculated for a perfect fluid, the flow would then be $q_{v1} = sV_1 = \sqrt{2gh};\ s = 1,26.10^{-3}\,m^2$; $V_1 = 9,9\,ms^{-1}$.

2) We calculate the Reynolds number in the horizontal tube using the velocity determined in 1. The result is indicative as in any case for a real fluid, the velocity really observed will be lower than V_P. This number is calculated using the kinematic viscosity ν.

μ, expressed in centipoises (therefore, in the cgs system), shows dynamic viscosity. The kinematic viscosity is deduced easily:

$$\mu = 1\,cps = 10^{-3}\,Pl \quad ; \quad \nu = \frac{\mu}{\rho} = \frac{10^{-3}}{10^{3}} = 10^{-6}\ stokes$$

Here, we used the following relation (very important to know): $1\,Pl = 10\,poises$

The Reynolds number is $R_{D1} = \frac{V_1 d}{\nu} = \frac{9,9.4.10^{-2}}{10^{-6}} = 3,96.10^{5}$

which shows that the flow can be assumed to be turbulent at this stage.

3) We determine the altitudes (cotes) z on a vertical axis with an upward direction; therefore, the origin is chosen on the axis of the horizontal tube. The head is defined by $H = \frac{V^2}{2g} + \frac{p_G}{\varpi}$, where p_G is the generating pressure $p_G = p + \rho g z$

and ϖ is the unit weight $\varpi = \rho g$. The head H_1 at the free surface of reservoir is:

$$H_1 = 0 + \frac{p_a + \rho g H}{\varpi} \qquad [5.39]$$

The head H_2 at the orifice of the horizontal tube is:

$$H_2 = \frac{V_S^2}{2g} + \frac{p_a + \rho g(0)}{\varpi} \qquad [5.40]$$

Seeing as the fluid is real, a loss of head is located exclusively along the length L of the tube (we neglect the effects of the section change coefficients). Using the Darcy formula, the loss of head ΔH between the two extremities of the tube is given by:

$$\Delta H = J\,L \quad ; \quad J = \psi \frac{V^2}{2\,g\,D} \qquad [5.41]$$

where J is the linear loss of head and ψ is the linear head loss coefficient. For the sake of simplicity, we shall choose for the determination of the linear head loss the expression suggested by Blasius:

$$R_D < 10^5 \; ; \; \psi = \frac{0,316}{R_D^{0,25}}$$

In all rigor, the initially estimated Reynolds number was significantly greater than 10^5, and as a result, the following expression should be used:

$$R_D > 10^5 \; ; \; \psi = 0,004 + \frac{0,25}{R_D^{0,25}}$$

However, the expected velocity, V_2, is smaller than V_1 and the Reynolds number should be smaller than 10^5, which we will verify later.

The loss of head is therefore equal to

$$\Delta H = \psi \frac{V^2}{2gd} L = \frac{0,316}{\left(\frac{V_2 d}{\nu}\right)^{0,25}} \frac{V_2^2}{2gd} L = \frac{0,316.\nu^{0,25}}{2gd^{1,25}} V_2^{1,75} L \quad [5.42]$$

$$\Delta H = \alpha V_2^{1,75} \quad [5.43]$$

$$\alpha = \frac{0,316.\nu^{0,25}}{2gd^{1,25}} L = 0,5694$$

which is a function of V_2, a real observed velocity, and which is the unknown variable.

H_1, H_2 *and* ΔH are therefore linked by:

$$H_1 = H_2 + \Delta H \quad [5.44]$$

with

$$H_1 = 0 + \frac{p_a + \rho g H}{\varpi} = \frac{p_a}{\varpi} + H \quad [5.45]$$

$$H_2 = \frac{V_2^2}{2g} + \frac{p_a}{\varpi} \quad [5.46]$$

$$5,09.10^{-2}\ V_2^2 + 0,5694\ V_2^{1,75} - 5 = 0$$

From this, we can deduce V_2.

An approximate solution can be obtained by writing $V_2^2 \approx V_2^{1,75}$.

The value of V_2 is intermediate between the values of V_{21} and V_{22}, which are the solutions of:

$$\left(5,09.10^{-2} + 0,5694\right) V_{21}^2 - 5 = 0$$

and

$$\left(5{,}09.10^{-2}+0{,}5694\right)V_{22}^{1{,}75}-5=0$$

This leads to:

$$V_{21}=2{,}84\,ms^{-1}\;;\;V_{22}=3{,}29ms^{-1}$$

The resulting flow rate is then contained between:

$$q_{V21}=sV_{21}=3{,}58.10^{-3}\,m^3s^{-1}\text{ and }q_{V22}=sV_{22}=4{,}14.10^{-3}\,m^3s^{-1}$$

We verify that the maximal Reynolds number

$$R_{D2}=\frac{V_{22}\,d}{\nu}=\frac{3{,}29.4.10^{-2}}{10^{-6}}=1{,}31.10^5$$

is indeed showing a turbulent flow, with a Reynolds number close to 10^5. We can get an exact solution to the equation by using a solver, available on many calculators. This would result in $V_2=3{,}248\,ms^{-1}$, which is well located within the previous values. The flow comes out as $q_{V22}=sV_{22}=4{,}09.10^{-3}\,m^3s^{-1}$.

The loss of head, which is a loss of mechanical energy through friction, can be found in flows that are not located within an inertial frame. We shall see an example of this later.

EXAMPLE 5.6 (The dental casting machine).–

The casting of some metallic dental prostheses is done using an *"dental casting machine"*. A horizontal rod ABD spins at an angular velocity ω around a vertical axis Dx. The branch DA carries a device M made up of the following:

– a pot C containing the molten metal to inject;

– a mold M_0 containing the cast of the prosthesis;

– a channel CM_0 linking the pot to the mold.

The machine is said to be "electronic" as the pot is heated by microwaves.

A counterweight C_P balances the systems over part DB of the rod. The channel CA has a length L and a diameter d. The molten metal has a density ρ and a kinematic viscosity ν. We assume that atmospheric pressure p_a governs C and M_0. The diameters of C and M_0 are negligible compared to L. The effects of gravity are neglected.

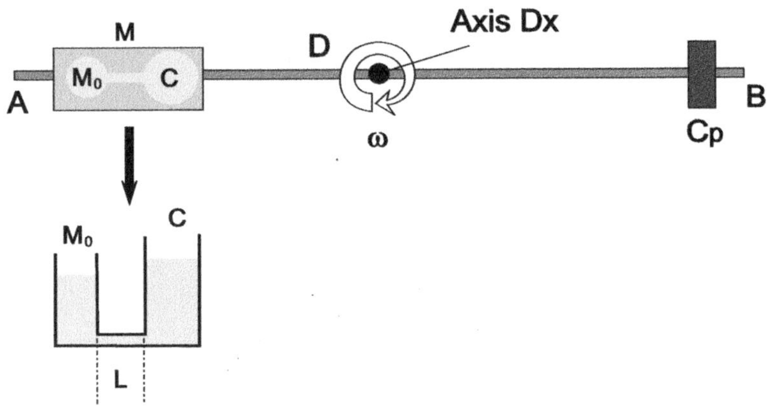

Figure 5.6. *The electronic casting machine*

The system is placed within a framework using an axis $M_0 z$ carried by AB and directed from A to B.

Figure 5.7. *The reference frame*

Once the metal has been melted in C, the system is made to rotate at an angular velocity of ω.

1) Give the expression of the filling time t_R of the mold M_0. The contraction coefficient of the jet in M_0 is taken as being equal to 1. In the channel and in the mold, we assume the molten metal behaves like a perfect fluid.

2) Molding a gold crown

Density of the gold $\rho = 19300\, kg.m^{-3}$

Viscosity of the gold $\nu = 1\, stokes$

$L = 5\, cm$, $d = 1\, mm$, $OM = 30\, cm$

Three grams of gold are required to make a crown. In practice, yellow gold used in dental prostheses is an alloy of 50–90% gold and other metals (silver, palladium, platinum, etc.). Calculate the filling time t_R of the mold for a rotation speed of $\omega = 240\, tours.mn^{-1}$.

Solution:

1) To know the filling speed, we must first know the volume flow $q_V = SV$ in the channel linking the reservoir to the mold. This involves searching for the velocity V in this channel.

First, it is important to reflect on the physics involved in this problem. The flow-generating device is located in a non-inertial framework. Indeed, the setup comprising the channel and the mold is in rotation at an angular velocity of ω in relation to the terrestrial reference frame. Inertia forces appear in the framework whose axis is M_{0z} . Seeing as the molding device is small compared to the rotation radius DM, we consider these inertia forces to be uniform over the molding device. These forces are directed from D toward M, which is in the opposite direction, and their intensity per unit of mass is $F_I = DM\,\omega^2$. Furthermore, we neglect the terrestrial forces of gravity. The force of intensity $F_I = DM\,\omega^2$ derives from a potential "per unit of mass" ϕ such that (this is in a right-angle corner Ox, Oy, Oz , with Oz being the only useful one here):

$$\vec{F}_I = (0, 0, DM\,\omega^2) = -\, Grad\,\phi \qquad [5.47]$$

This results in the following expression of ϕ :

$$\phi = DM\,\omega^2 z = g'z \qquad [5.48]$$

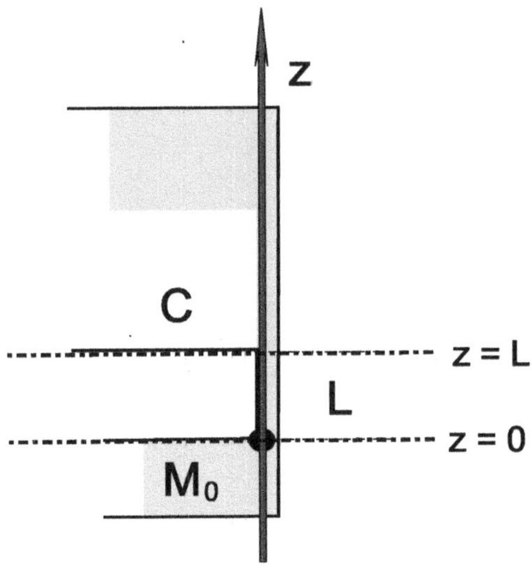

Figure 5.8. *The flow of the gold*

The problem can be seen as a vertical flow between C and M_0 on a "planet" where gravity would be equal to $g' = DM\,\omega^2$. For this, we need $g' >> g$, which we will justify in the applications.

In this context, a head can be defined from the volume forces through:

$$H = \frac{V^2}{2g'} + \frac{p}{\rho g'} + z \qquad [5.49]$$

The head has the values respectively in C and M_0 of

$$H_C = \frac{0}{2g'} + \frac{p_a}{\rho g'} + L \qquad [5.50]$$

$$H_{M_0} = \frac{0}{2g'} + \frac{p_a}{\rho g'} + 0 \qquad [5.51]$$

Viscosity only intervenes between C and M_0. This causes a localized loss of head in the channel CM_0. We would presume the flow is slow (it is a molten metal) and assume that the flow is laminar.

In this case, the loss of head is calculated through

$$J = \psi \frac{V^2}{2g'd} \quad ; \quad \Delta H = J\,L \qquad [5.52]$$

with $\psi = \frac{64}{R_D} = \frac{64 * \nu}{V\,d}$ [5.53]

By showing V:

$$\Delta H = \frac{64 * \nu}{V\,d} * \frac{V^2}{2g'd} L = \alpha V \qquad [5.54]$$

$$\alpha = \frac{64 * \nu}{2g'd^2} = \frac{32 * \nu L}{DM\,\omega^2 d^2} \qquad [5.55]$$

The relation between H_C and H_{M_0} is then simply:

$$H_C = H_{M_0} + \Delta H \qquad [5.56]$$

In terms of energies, this expression shows that between C and M_0 and along the axis $M_0\,z$, the fluid has lost, per unit of volume, an amount of energy equal to $\rho\,g' = \rho\,OM_0\,\omega^2$, which has been transformed into heat by friction in the channel.

This provides us with the equation that allows us to calculate V:

$$\frac{p_a}{\rho g'} + L = \frac{p_a}{\rho g'} + \alpha V \qquad [5.57]$$

There is $V = \frac{L}{\alpha}$ and $\alpha = \frac{32 * \nu L}{DM\,\omega^2 d^2}$. On the axis $M_0\,z$, point C has a cote of L. Furthermore, C M_0 are at atmospheric pressure.

2) We can calculate the numerical values of α and V since $q_V = \pi \frac{d^2}{4} V$.

ν expressed in Stokes is therefore a kinematic viscosity, expressed in the cgs system. It must be converted into $myriastokes$, or, as is more common, into $m^2.s^{-1}$: $\nu = 10^{-4}\, m^2.s^{-1}$.

Furthermore, $\omega = 240\, turns.mn^{-1} = 4\, turns.s^{-1} = 6 * 2\pi = 8\pi\, Rd.s^{-1}$:

$$\alpha = \frac{32 * \nu L}{DM\, \omega^2 d^2} = \frac{32.10^{-4} * 5.10^{-2}}{0,3 * (8\pi)^2 * (1.10^{-3})^2} = 0,844$$

$$V = \frac{L}{\alpha} = 5,92.10^{-2}\, m.s^{-1} = 5,92\, cm.s^{-1}$$

The volume flow becomes:

$$q_V = \pi \frac{d^2}{4} V = 7,85.10^{-7} * 5,92.10^{-2} = 4,65.10^{-8}\, m^3.s^{-1}$$

3 g of gold represents a volume of $V_G = \dfrac{3.10^{-3}}{19300} = 1,554.10^{-7}\, m^3$

With a flow q_V = $5,3.10^{-5}\, m^3.s^{-1}$, the filling time is:

$$t_R = \frac{V_G}{q_V} = \frac{1,554.10^{-7}}{4,65.10^{-8}} = 3,34\, s$$

By comparing g and g', we get:

$$g' = DM\, \omega^2 = 0,3 * (8\pi)^2 = 189,5$$

There is $\dfrac{g'}{g} = 19,3$

which justifies neglecting the forces of gravity. Thus, the changes in the level of the gold in the mold do not influence the instantaneous flow. To get a better idea, if the gold was contained in a single closed vase, the pressure difference between the entrance and the exit of the channel would be:

$$\Delta p_{CM_0} = \rho\, DM\, \omega^2\, L = 19300 * 0,3 * (8\pi)^2 * 5.10^{-2} = 1,83.10^5\, Pa = 1,8\, bar$$

6

Calculation of Thrust and Propulsion

6.1. Introduction

The emission of a jet from a moving solid and the interaction of a jet with a fixed or mobile wall form the basis of many practical problems, and are involved in the important technical area of propulsion. In this field, the *Euler theorems* are key in resolution. While two of these theorems exist, it is mainly the first that is used, and as such is often referred to as *the* Euler theorem, with the second tending to be forgotten (which is not a good practice!). We shall not adopt this habit here, and at the end of the chapter we shall look at the uses of the second Euler theorem. "The" Euler theorem can be applied both to perfect fluids and real fluids. A whole chapter is therefore dedicated here to some of the simple applications of this theorem. While the theorem can be used in a far larger scope of application, here we shall focus on problems involving the notion of propulsion, or jet–wall interaction.

6.2. Euler's theorem and proof

In this chapter, we will look at two characteristic situations:

– *The case of jet propulsion systems.* Two questions are asked in this case:

- What is the value of the thrust $\vec{P}$ produced by a device?
- What is the yield of the propulsion η?

– *The case of the impact of a jet on a wall*

Both types of problem can interact together in certain technological devices, such as an engine (elementary study of propulsion, elementary study of the turbojet turbine, etc.).

6.2.1. *Euler's first theorem and proof*

What is expressed in Euler's theorem is not always very well understood. For this reason, we have included its proof in its elementary form.

6.2.1.1. *Hypothesis of Euler's theorem*

This theorem requires only one hypothesis: we are in a framework where the flow is stationary. It is not possible to find such a reference for all flows, and in such a case the theorem cannot be written.

Let us consider an elementary streamtube. We look at the fluid that, at the moment *t*, is contained between two surfaces dS_1 and dS_2 against which the streamtube is supported.

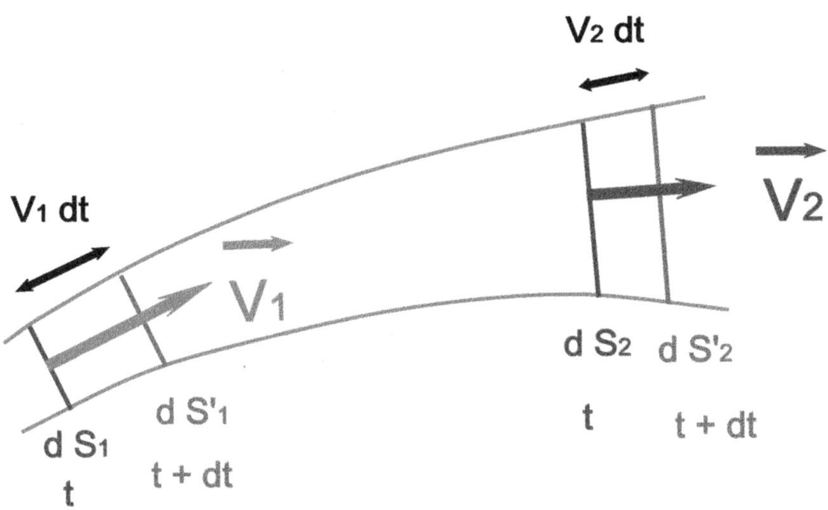

Figure 6.1. *Euler's theorem applied to an elementary streamtube*

The principle of continuity allows us to write the expression of the mass flow that is conserved the length of the streamtube:

$$dq_m = \rho_1 V_1 dS_1 = \rho_2 V_2 dS_2 \qquad [6.1]$$

It is important to note that, with regard to the unique hypothesis adopted, there is no reason for V_1 and V_2, as well as ρ_1 and ρ_2, respectively, to be the same in one section of the streamtube as in another.

The fluid contained in the streamtube receives various forces. It is the resultant force on the fluid contained at the moment t between dS_1 and dS_2 that we shall look at now.

It should be noted that at no moment do we look to make a distinction between the volume forces and surface forces. In any application of Euler's theorem, this will imply obtaining further information on these forces.

We write the fundamental principle of dynamics for this fluid between dS_1 and dS_2. We write the principle in the form of a theorem on the quantity of movement. More exactly, we write that during the time dt the temporal derivative of the quantity of movement in the fluid in question (or in other words, the change "per unit of time" of this quantity) is equal to the resulting force from the forces applied to it during this time.

This is obviously a vectorial relation, to take into account the changes in the direction of the fluid throughout its whole trajectory.

During the time dt, the fluid, which remains in the current tube, goes from a volume contained between dS_1 and dS_2 to a volume contained between dS'_1 and dS'_2. The surfaces dS'_1 and dS'_2 are respectively at a distance of $V_1\,dt$ and $V_2\,dt$ dS'_2 from dS_1 and dS_2.

From a Eulerian perspective, the quantity of movement of the fluid particles contained between dS'_1 and dS_2 does not change over time, although these are not the same fluid particles that "carry" the quantity of movement in each fixed point.

The only variation during dt of the quantity of movement of the fluid contained between dS_1 and dS_2 lies in:

a) the quantity of movement "acquired" during dt, which is that of the fluid contained between dS_2 and dS'_2, or in other words, the product of the volume, density and local velocity: $\rho_2\,V_2\,dS_2\,dt\,\vec{V}_2$.

b) the quantity of movement "lost" during dt, which is that of the fluid contained between dS_1 and dS'_1, or in other words, the product of the volume, density and local velocity:

$$\rho_1\,V_1\,dS_1\,dt\,\vec{V}_1 \qquad [6.2]$$

The resulting force $\vec{F} = \vec{F}_V + \vec{F}_S$ from the forces applied to this fluid is therefore equal to the ratio of this summary of quantity of movement to dt:

$$\vec{F} = \frac{\rho_2 V_2 dS_2 dt \vec{V}_2 - \rho_1 V_1 dS_1 dt \vec{V}_1}{dt} \quad [6.3]$$

In this expression, we can show the mass flow rate:

$$dq_m = \rho_1 V_1 dS_1 = \rho_2 V_2 dS_2 \quad [6.4]$$

and get rid of dt. This gives us:

$$\vec{F} = dq_m \left(\vec{V}_2 - \vec{V}_1\right) \quad [6.5]$$

which is, once again, just a form of writing the fundamental principle of dynamics. This expression is translated by saying that the resulting force from those applied to the fluid between dS_1 and dS_2 is equal to the difference between an exiting *quantity of movement flow* rate $d\,q_m\left(\vec{V}_2\right)$ and an entering *quantity of movement flow rate* $d\,q_m\left(\vec{V}_1\right)$.

The notion of quantity of movement flow rate can sometimes be surprising for readers and auditors. Knowing that this is a vectorial quantity, there is no reason to be surprised: the fluid enters or exits from each elementary surface of S with a velocity vector $\vec{V}$. The corresponding mass $d^2 m = dq_m\, dt$ "carries" a quantity of movement $\left(d^2 m\right)\vec{V}$ (which is a vector). The notion of flow rate, of the quantity of movement going through S per unit of time need not be surprising.

6.2.1.2. *Proof of Euler's theorem in a finite volume*

This proof can be extended to a finite volume limited by a given surface S, fixed in relation to our framework where the flow is permanent. The flow that goes through this surface can also be divided into elementary streamtubes, and then all the streamtube forces and flow rates can be summed from the theorem proven previously.

6.2.1.3. *Euler's two theorems: writing in vectorial form*

In general, Euler's theorem is written in vectorial form. Here we give this expression, while remembering that this form of writing may not be a familiar one for all readers of this work. With regard to Euler's first theorem, equation [6.6] only formalizes what was written in section 6.2.1.2. There are two Euler's theorems:

Euler's first theorem:

$$\iint_S \vec{V}\, dq_m = \iiint_D \vec{F}_V\, d\omega + \iint_S \vec{F}_S\, dS \quad [6.6]$$

Euler's second theorem:

$$\iint_S \vec{r}\times\vec{V}\, dq_m = \vec{r}\times\left[\iiint_D \vec{F}_V\, d\omega + \iint_S \vec{F}_S\, dS\right] \quad [6.7]$$

Euler's first theorem is the one we have just proven. The first equation formalizes the fact that the balance of the quantity of movement flow rates through a finite closed surface S (left term) is equal to the resulting force applied to the fluid contained within this surface. The term on the right-hand side of equation [6.6] makes a distinction in writing between surface forces and volume forces.

Euler's second theorem does not always attract readers' – or sometimes authors' – attention. It states that the balance of *moment* of quantity of movement flow rate is equal to the *resulting moment* of the applied forces. This goes back to an issue encountered in the elementary classes on the physics of solids: a resulting force of the forces applied to a solid equal to zero does not mean that there is no movement if the resulting moment of these forces is not equal to zero.

It must be noted that many textbooks, even from very good authors, disregard the *proof* of this theorem. In a way, Euler's first and second theorems are the counterpart of the theorem on the center of mass and the theorem on the kinematic moment in solid mechanics. We shall make these theorems "come alive", the expression of which through [6.6] and [6.7] can seem abstract for some readers, by applying it to the problems of jet propulsion and of thrust.

The first application directly involves aeronautics. It is important to note that while the example in the following paragraph involves jet planes, the jet propulsion phenomenon is also relevant for "propeller" airplanes. Indeed, propeller propulsion is also a form of propulsion by reaction. Froude's theory is an illustration of this. The second application mostly focuses on issues seen in turbo machines, with calculations that are far more complex than the more simple examples present later on.

6.2.1.4. *Summary*

Before showing these examples of application, let us recall that:

a) Euler's theorem implies being within a framework where the flow is permanent. Such a framework must therefore be possible.

b) Euler's theorem allows us to determine the resulting force of those applied to an immobile fluid domain in this framework. The only piece of information required is knowledge of the kinematics present on the surface that marks the limit of this domain (and for compressible fluids, knowledge of the distribution of the densities on this surface).

c) However, only one resulting force can be obtained from the volume forces and surface forces. One additional piece of information is vital for distinguishing these forces (which is usually a prerequisite for the problem in question).

d) There is a second Euler's theorem that relates to the moments of the forces applied to a volume of fluid.

e) Finally, as this is very important, it can be shown that if a solid is present inside the surface S, the resulting force of the forces calculated using Euler's theorem includes the forces applied to this solid part.

The proof calls on the law of action and reaction; we shall not reproduce this here. It is important to understand that *only the forces applied to the solid part included in S are considered*: thus, if there are forces of gravity, only the solid mass present in S appears in the calculation.

6.3. Thrust of a jet propulsion system, and propulsive efficiency

The expression of thrust is useful for the resolution of many of the problems presented here; as such, we shall make a generalizable presentation of it in the following examples.

6.3.1. *Calculation of the thrust of an "airplane engine"*

Here we provide a general form of reasoning, resulting in a simple representation of the thrust and propulsive efficiency. In some of the textbooks, the presentation of the propulsive efficiency is not quite complete. This can mean that the relation between the initial energy provided and the energy consumed by the propulsion is provided, but no mention is made of where the "lost" energy goes. Here we shall fill this in.

We shall need to use Euler's first theorem by projection onto a unique axis, usually denoted by Ox.

In a normal jet system (engine), the axis of the engine is enough for this projection. In our reasoning, we orientate Ox in the direction of the exiting matter.

A distinction must be made between:

a) the quantity of matter going through the engine per unit of time;

b) the velocity vectors of this matter as it enters or exits.

A summary of the quantity of movement entering and exiting per unit of time is made, called the "quantity of movement flow rate". However, this quantity of movement is vectorial.

A flow of matter perpendicular to the axis at Ox corresponds not only to a finite mass flow rate q_m, but also to a "projected" quantity of movement flow that is equal to zero, as the support of the velocity, which is also the support of the quantity of movement, is perpendicular to the axis Ox.

Let us take the case of a classic jet engine (the reasoning is just as a valid for a propeller, which is also a "reaction" system).

An airflow of q_m goes through this jet engine.

The jet engine is installed on an airplane moving at a constant velocity V.

We are in the framework of the jet engine, which is a Galilean framework, as V is constant.

We already noted in Chapter 4 that a mobile object moving at a constant velocity could constitute a Galilean framework. In this case, there is no need to look for forces of inertia in the summary of the volume.

The thermal energy brought to the engine increases the velocity of the air exiting it in relation to the entrance velocity.

The exact velocity of the gas on entry into the engine is unknown. We know that far ahead of the airplane the flow q_m is moving at a velocity V ($+V$ in the direction of our axis) related to the airplane.

This is the source of a common error; the statement comes from the same reasoning seen in Chapter 4 in the problems of filling when using the Bernoulli theorem.

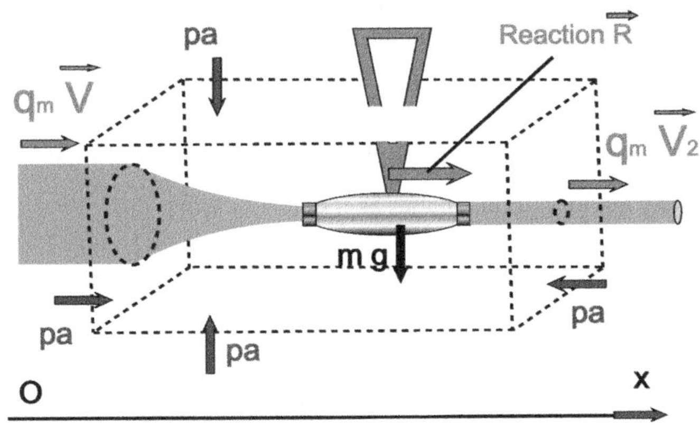

Figure 6.2. *Application of Euler's theorem to a jet propulsion engine*

For the application of Euler's theorem, our closed reference surface will be of greater dimension than the engine's outer casing.

In this surface, which encases a solid surface, but which is immobile in relation to the reference frame used, the flow q_m enters at a uniform velocity $\vec{V}$ of the modulus V and exits at a velocity $\vec{V}_S$ of the modulus V_S in the direction of the axis, which we assume to be uniform for the sake of simplicity.

NOTE: The case of the dual flow jet engine requires the development of a more complex analysis.

The projection of the term $\iint_S \vec{V}\, d\,q_m$ on the axis Ox then takes on a very simple form:

$$\mathrm{Pr}\,oj_{Ox}\left[\iint_S \vec{V}\, dq_m\right] = q_m\left(V_S - V\right) \qquad [6.8]$$

V has a minus sign in the brackets as the normal to the surface elements through which the fluid "enters" are turned from the inside to the outside of the closed surface. It is therefore the q_m factor V that is negative.

Let us consider an engine whose movement is horizontal.

The projections of the forces applied to all that is in the reference surface (including the solids) are easy to write.

The pressure forces are made up on the front and back and top and bottom faces of the reference surface.

In terms of the volume forces:

a) the forces of gravity have a support that is perpendicular to Ox and whose projection is therefore equal to zero;

b) the only "remaining" volume force is carried by the axis Ox. This is the reaction of the airplane on the wing. It is constant in reaction problems.

It is important to note that this is a force applied to the solid present within the reference surface. Hence, it is important to take on board point (e) of 6.2.1.4 and to be conscious of the expression of Euler's theorem when in the presence of the solid surface in this reference surface.

It is not in fact the thrust of the jet engine that is being calculated, but rather the value of the reaction force of the jet that goes against this thrust.

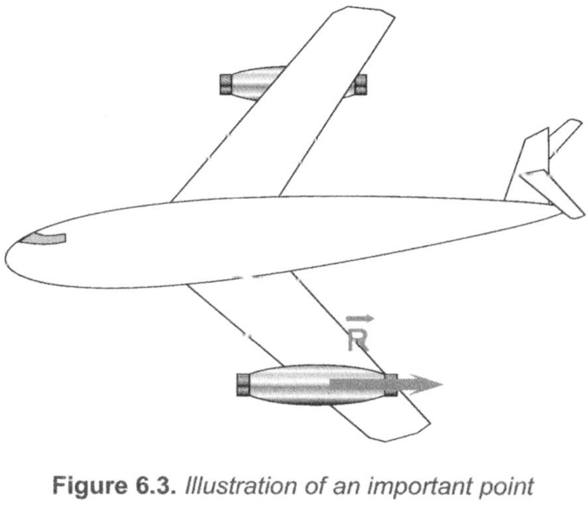

Figure 6.3. *Illustration of an important point*

We do not determine the thrust of the engine, but rather the reaction $\vec{R}$ applied by the airplane onto the engine to "stop it accelerating".

Moreover, two cases can be found:

a) the case where the plane has a constant velocity, and the reference frame is inertial: the thrust is then equal and opposed to the reaction of the wing of the airplane.

b) an extension to the case where the movement of the airplane is accelerated. The form found in the present model is then still used, although the reference frame is no longer Galilean for the engine. This approximation is acceptable for moderate accelerations.

Let us call this force $\vec{R} = -\vec{P}$, equal and opposed and therefore equal in intensity to the thrust. The term $\iiint_D \vec{F}_V \, d\omega + \iint_S \vec{F}_S \, dS$, projected on Ox then becomes

$$\mathrm{Proj}_{Ox}\left[\iiint_D \vec{F}_V \, d\omega + \iint_S \vec{F}_S \, dS\right] = R \qquad [6.9]$$

Finally, this gives the expression of the projection of Euler's theorem, which provides us with P:

$$q_m \left(V_S - V\right) = R \qquad [6.10]$$

The intensity of the thrust can be deduced simply, like $R = P$

$$P = q_m \left(V_S - V\right) \qquad [6.11]$$

NOTE.– This expression assumes that the gases enter and exit the engine axially (see Example 6.6).

6.3.2. *Calculation of the propulsive efficiency*

The thermodynamic efficiency of the engine links the quantity of work provided that is usable for propulsion and the amount of heat created by the combusted kerosene. In the case of a jet engine, this yield is the product of a "thermal efficiency", which is the ratio of the kinetic energy transferred to the air in the engine to that of the "propulsive efficiency", whose analysis is more delicate, and is studied here. Here we consider that the first efficiency, the thermal efficiency, is very close to one, which is not far from reality.

This propulsion efficiency η compares:

a) the work carried out per unit of time by the propulsion force, product of the force of thrust and the distance traveled by the airplane per unit of time. This work is a power:

$$W = P * V = q_m (V_S - V) V \qquad [6.12]$$

b) the energy consumed by the motor. This energy has carried per unit of time the kinetic energy of the incident air $\frac{V^2}{2}$ per unit of mass to the kinetic exit energy $\frac{V_S^2}{2}$ per unit of mass. The energy consumed per unit of time, E, which affects a mass of air (q_m) is therefore equal to:

$$E = q_m \left(\frac{V_S^2}{2} - \frac{V^2}{2} \right) \qquad [6.13]$$

The propulsive efficiency η is therefore:

$$\eta = \frac{W}{E} = \frac{q_m (V_S - V) V}{q_m \left(\frac{V_S^2}{2} - \frac{V^2}{2} \right)} \qquad [6.14]$$

After simplification:

$$\eta = \frac{2}{1 + \frac{V_S}{V}} \qquad [6.15]$$

The optimization of this efficiency must verify two contradictory constraints:

a) increase the difference between the entering and exiting velocities $(V_S - V)$ to increase thrust;

b) reduce this difference to decrease the ratio $\frac{V_S}{V}$ and increase the propulsive efficiency.

These constraints are tied together in the *dual flow jet engine*, where q_m is increased to increase thrust and $(V_S - V)$ is reduced.

The issue remains: an efficiency of less than one means a "loss of energy". *Where does the lost energy go*?

To answer this question, we must be in the reference frame of the ground: before the airplane passes through, the air is immobile and its kinetic energy, in the frame of the ground, is equal to zero; after its passage, the air exiting the engine has a velocity $(V_S - V)$.

For a mass of air q_m, an amount of energy is "lost" in the air kinetically, which is expressed by [6.13]:

$$E_{air} = q_m \frac{(V_S - V)^2}{2} \quad [6.16]$$

Let us verify that E_{air} is indeed the energy lost:

$$E - W = q_m \left[\left(\frac{V_S^2}{2} - \frac{V^2}{2} \right) - (V_S - V)V \right] \quad [6.17]$$

$$\left(\frac{V_S^2}{2} - \frac{V^2}{2} \right) - (V_S - V)V = \frac{1}{2}\left(V_s^2 - V^2 - 2V_sV + 2V^2\right) = \frac{V_s^2 + V^2 - 2V_sV}{2} \quad [6.18]$$

And we find again:

$$E - W = q_m \frac{(V_S - V)^2}{2} = E_{air} \quad [6.19]$$

The kinetic energy E_{air} is next thermalized by viscous dissipation in the surrounding air. From a thermodynamics point of view, this goes back to the notion of heat lost to a cold source. The engine is indeed a thermal motor that functions between a hot source (fed by the kerosene supplied) and a cold source, which is the atmosphere in which the airplane is traveling.

6.3.3. *Calculation of the thrust of a rocket engine*

In the case of a mobile object such as a rocket, the jet providing propulsion is generated by the mobile object itself. The fuel is contained in the engine. There is therefore no mass flow of air entering the system. It is important to note that the flow of the exiting fluid is not usually a flow of air.

The above reasoning remains valid, and the result obtained for the thrust remains the same. Because of a lack of air entry, the velocity of the projectile disappears from the expression of the thrust.

A rocket moves at a velocity V and has a jet whose mass flow is equal to q_m of a fluid whose velocity relative to the rocket is V_S.

The thrust of a rocket is therefore:

$$P = q_m V_S \quad [6.20]$$

The propulsive efficiency is calculated by evaluating the energy

$$E_C = q_m \frac{V_S^2}{2} \quad [6.21]$$

provided per unit of time to the exiting gas and by comparing it to the energy spent in the propulsion, which is a unit of time:

$$W = PV = q_m V_S V \quad [6.22]$$

This results in the propulsive efficiency η:

$$\eta = \frac{W}{E_C} = \frac{q_m V_S V}{q_m \frac{V_S^2}{2}} = 2\frac{V}{V_S} \quad [6.23]$$

6.3.4. *Some applications of Euler's theorem to jet propulsion*

The following examples involve systems driven by a "jet engine". They make extensive use of the results established previously.

EXAMPLE 6.1 (Simplified model of a single-flow jet engine).–

An airplane is equipped with a single jet engine. The air exits the engine at a temperature of T_2, through a section Ox. A mass flow of air $q_m = 70\,kg.s^{-1}$ goes through the engine.

We assume that the Boyle Mariotte law applies to the air. The airplane flies at an altitude of $10000\,m$. The surrounding temperature is $T_1 = 223\,K$. The density of the air is then $\rho_1 = 0,1\, kg.m^{-3}$. The gases exit at a level S_2 at a temperature of $T_{S2} = 700\, K$.

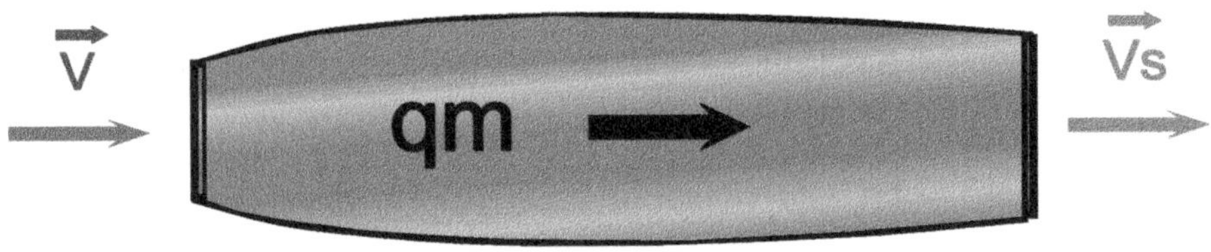

Figure 6.4. *Single-flow jet engine*

The airplane flies at a velocity $V = 700\, km.hr^{-1}$, with a rectilinear and horizontal trajectory.

Throughout the problem, we assume that the aerodynamic friction of the flight can be reduced to a single resulting force $\vec{F}(V)$, where V is the velocity of the airplane in relation to the ground. This reaction is collinear to the trajectory of the airplane and obviously opposed to its movement. We have $F = k\,V^2$, where k is a constant.

1) Thrust of the engine.

1.1) Find the value of V_S, exit velocity of the gases relative to the engine.

1.2) Recall the expression of the thrust of the engine P. Calculate it, and express it in Newton and in tons of force.

1.3) Deduce the value of k.

2) Thrust reverser.

The airplane is landing. It rolls along the landing strip at $V' = 300\, km.hr^{-1}$. Two deflectors are opened behind the jet engine (see Figure 6.5), called thrust reversers, which can be represented by two panels making an angle of 60° with the horizon. For the sake of simplicity, we assume that the panels cause a simple deflection, and the modulus V_2 of the exit velocity of the gases is conserved upon exiting the deflector. We keep the values found in question (1) for q_m and V_2. What is the braking force P_R applied on the airplane by the thrust reverser?

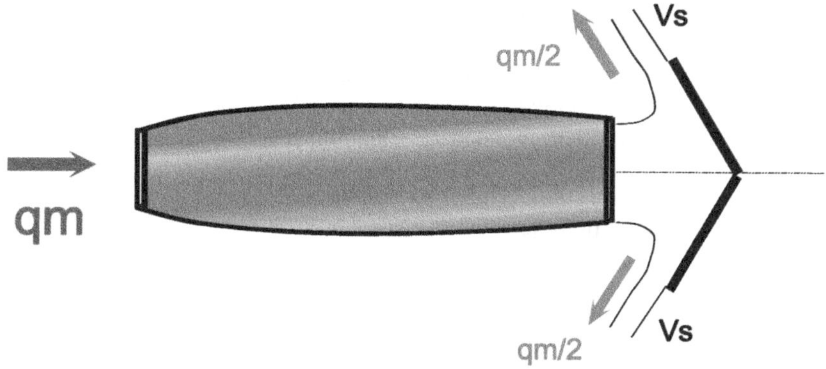

Figure 6.5. *Thrust reversers*

Solution:

1.1) Considering that atmospheric pressure governs the entrance and exit of the engine, the density of the exiting air is:

$$\rho_2 = \rho_1 \frac{T_2}{T_1} = 3{,}186.10^{-2}\ kg.m^{-3} \quad [6.24]$$

V_S can then be deduced from the mass flow $q_m = \rho_2\, S_2\, V_S$ [6.24]

$$V_S = \frac{70}{3{,}186.10^{-2} * 4} = 549{,}3\, m.s^{-1} \quad [6.25]$$

1.2) Going back to the results proven in the Introduction:

$$P = q_m\left(V_S - V\right) \quad [6.26]$$

$$V = \frac{700\,000}{3600} = 194{,}4\, m.s^{-1} \quad [6.27]$$

$$P = 70 * (549{,}3 - 194{,}4) = 2{,}48.10^4\ N \quad [6.28]$$

Remember that a kilogram of force is the intensity of the force equal to the weight of a mass of 1 kg; therefore $1\,kgf = 9{,}81\,N$

$$P = 2{,}48.10^4\ N = 2532\,kgf = 2{,}532\,tf \quad [6.29]$$

The engine is said to have a thrust of 2.5 tons

1.3) When the velocity is constant (and therefore the value of the acceleration is equal to zero), the forces applied to the airplane balance out. At this moment, it is important to remember that in such a situation, Euler's theorem allows us to calculate F and not P.

In terms of intensity, there is:

$$P = F = kV^2 \quad [6.30]$$

Therefore: $k = \frac{P}{V^2} = \frac{2,48.10^4}{194,4} = 0,656\, Nm^{-2}s^2$ [6.31]

2) We use the axis Ox of the engine, pointed in the inverse direction of the movement. The force $\vec{P}_R$ that we are looking for is equal and opposite to the force $\vec{F}_R$ applied by the airplane on the engine. By projection onto the axis Ox of the engine (from left to right in Figure 6.5), Euler's theorem is written as:

$$F_R = \frac{q_m(-V_S \cos\alpha)}{2} + \frac{q_m(-V_S \cos\alpha)}{2} - q_m V' \quad [6.32]$$

$$V' = \frac{300\,000}{3600} = 83,33\, ms^{-1} \quad [6.33]$$

$\cos\alpha$ here expresses the projection of the velocity V_S on Ox; $\alpha = 60°$, therefore $\cos\alpha = 0,5$. In the first two terms, q_m is positive as it is matter that exits, $V_s \cos\alpha$ has a minus sign, as it is the "negative component" that exits.

$\vec{F}_R$ is a force that has the same direction as the movement (its component F_R on Ox comes out negative). $\vec{P}_R$ is equal and opposed to it. The two forces are equal in intensity. In fact, the thrust reverser tends to decelerate the airplane. The framework of the engine is no longer rigorously inertial. Here we are in the situation where the deceleration is still not significant, hence the use of V; the calculation is a good estimation of the thrust reversal.

Hence, the thrust is linked to the airplane's reverser. In terms of intensity:

$$P_R = F_R = q_m \left(V_S \cos\alpha + V\right) \qquad [6.34]$$

$$P_R = 70 * \left(549,3 * 0,5 + 83,33\right) \qquad [6.35]$$

The thrust of the reverser is therefore:

$$P_R = 25\,057\,N \qquad [6.36]$$

EXAMPLE 6.2 (Simplified model of a dual-flow jet engine).–

An airplane equipped with a single-flow jet engine flies at a velocity of $V = 600\,km.hr^{-1}$. The single-flow reactor it is equipped with has an airflow $q_m = 40\,kg.s^{-1}$. The exit velocity of the gases is then $V_{S1} = 600\,m.s^{-1}$.

1) What is the thrust P of this engine and what is its propulsive efficiency η_1 ?

2) The airplane is equipped with a new engine made from the old engine but surrounded by a coaxial enclosure through which a gas flow of q'_m travels. A "dilution ratio" λ is defined by: $q'_m = \lambda q_m$. We choose $\lambda = 4$.

We arrange the primary jet engine and the secondary flow in such a way that the thrust P of the new engine is the same as the single-flow engine from question (1). Moreover, the new exit velocity of the gas V_{S2} is identical for the primary engine and the secondary flow.

NOTE: In practice, this condition is difficult to obtain. The primary flow goes through a combustion chamber, while the secondary flow is generated by a compressor. We will accept this slight glitch to simplify the question.

The problem here "idealizes" the data.

We keep the same value of $q_m = 40\,kg.s^{-1}$ for the flow in the primary engine.

2.1) What is the value of the velocity V_{S2} ?

2.2) Calculate the new propulsive efficiency η_2.

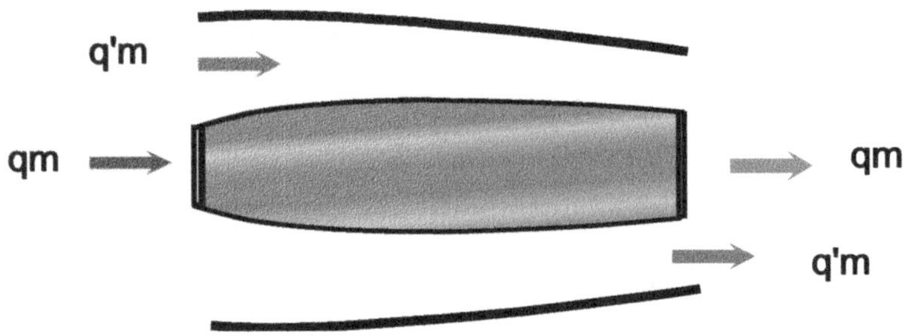

Figure 6.6. *Dual-flow jet engine*

Solution

1) Single flow

The thrust is written as

$$P = q_m (V_S - V) \tag{6.37}$$

The value of this thrust is:

$$V = \frac{600\,000}{3600} = 166,7\, m.s^{-1} \tag{6.38}$$

$$P = 40 * 600 - 166,7 = 1,733.10^4\ N = 1,77\, tf \tag{6.39}$$

The propulsion yield is calculated as:

$$\eta_1 = \frac{2}{1 + \frac{V_{S1}}{V}} \tag{6.40}$$

$$\eta_1 = \frac{2}{1 + \frac{600}{166,7}} = 0,435 \tag{6.41}$$

2) Dual flow

2.1) The new expression of the thrust is:

$$P = (q_m + q'_m)(V_{S2} - V) = (\lambda + 1) q_m\ (V_{S2} - V) \tag{6.42}$$

$$P = 5\, q_m \left(V_{S2} - V\right) \quad [6.43]$$

$$P = q_m \left(V_{S1} - V\right) \quad [6.44]$$

$$V_{S2} - V = \frac{V_{S1} - V}{5} \quad [6.45]$$

$$V_{S2} = \frac{600 - 166,7}{5} + 166,7 = 253,4\, m.s^{-1} \quad [6.46]$$

2.2) The new propulsion yield is:

$$\eta_2 = \frac{2}{1 + \dfrac{V_{S2}}{V}} \quad [6.47]$$

$$\eta_2 = \frac{2}{1 + \dfrac{253,4}{166,7}} = 0,794 \quad [6.48]$$

These orders of magnitude are realistic. The SE210, or "Caravelle", the first French civilian airplane (which left the Dassault factory in 1955), had a propulsion yield in the order of 0.3. In modern engines, whose dilution rates are often between 6 and 9 (they sometimes are as high as 15) yields of more than 0.7 are common.

EXAMPLE 6.3 (Jet propulsion of a catamaran).–

A catamaran is a boat with two hulls, which are usually parallel to each other. Various modes of propulsion can be used: other than the most traditional, the sail, water-jet propulsors (jet engines, in fact) are used in catamarans running along certain European naval paths.

Figure 6.7. *Picture of a catamaran and diagram of a jet propulsion system*

A diagram made available for travelers on a catamaran crossing Øresund, linking Copenhagen and Malmö states:

a) the ship is driven by two water jet propulsors (water turbines);

b) the flow rate of each jet is $q_m = 3000\, l.s^{-1}$ (this rate is determined in the reference frame of the jet generator);

c) the ship is equipped with two diesel engines whose total power is equal to P = 2 090 hp;

d) the maximal velocity of the ship is 32 knots;

Moreover, we assume that:

e) the two generators have the same jet flow rate, as well as the same jet ejection velocity. V_S is this exit velocity, determined within the reference frame of the catamaran.

f) all the energy of the diesel engines is used and distributed evenly between the two jet generators of the propulsors.

g) throughout the entire problem, the catamaran is going at its top velocity, and for that it uses the maximal amount of energy available from the diesel engines.

We recall that one horsepower is worth $750\,W$ and that a knot is the naval unit of speed, equal to a nautical mile per hour. A nautical mile is worth $1\,mile = 1852\,m$. The density of the water is $\rho = 100\,kg.m^{-3}$.

The water from the jets is collected from the front of the catamaran, along the axis of the movement.

Calculate:

1) The ejection velocity of the liquid V_S.

2) The thrust P of each of the jets, the total thrust of the catamaran.

3) The propulsive efficiency η of this propulsion.

Solution:

1) The power of each reactor is known

$$W = \frac{2090}{2} = 1045\,W \qquad [6.49]$$

This power is the work applied by the thrust P of the reactor during the unit of time. In 1 s, the catamaran covers V .

The thrust of a reactor is given by:

$$P = q_m \left(V_S - V\right) \qquad [6.50]$$

The power is therefore:

$$W = PV = q_m \left(V_S - V\right)V \qquad [6.51]$$

All the parameters are known except V_S, which is deduced from

$$W = 1045 * 750 = 7{,}84.10^5\, W \qquad [6.52]$$

$$q_m = 3000\, kg.s^{-1} \qquad [6.53]$$

$$V = \frac{32*1852}{3600} = 16{,}46\, m.s^{-1} \qquad [6.54]$$

$$W = 7{,}84.10^5 = 3000\left(V_S - 16{,}46\right)*16{,}46 \qquad [6.55]$$

$$V_S = 32{,}33\, m.s^{-1} \qquad [6.56]$$

2) The thrust of each of the jets is then:

$$P = q_m \left(V_S - V\right) = 3000\left(32{,}33 - 16{,}46\right) - 4{,}76.10^4\, N = 4{,}84\, tf \qquad [6.57]$$

The total thrust of the catamaran is twice this thrust:

$$P_T = 2 * P = 9{,}52.10^4\, N = 9{,}7\, tf \qquad [6.58]$$

3) The yield of this propulsion is written as:

$$\eta = \frac{2}{1 + \frac{V_S}{V}} \qquad [6.59]$$

$$\eta = \frac{2}{1 + \frac{32{,}33}{16{,}46}} = 0{,}6747 = 67{,}5\,\% \qquad [6.60]$$

The propulsive efficiency of a reactor is obviously also the propulsive efficiency of all of the elements of the two-engine system.

EXAMPLE 6.4.–

A baby's bath toy is a little boat in which a reservoir is constructed (see Figure 6.8) in the shape of a tube T, rectangular in section, whose sides are a and b; this reservoir empties itself into the bath through a horizontal jet that exits from a small horizontal tube of diameter d ($d^2 << ab$). The horizontal tube acts like a jet engine, with a thrust of P, thus driving the boat forward.

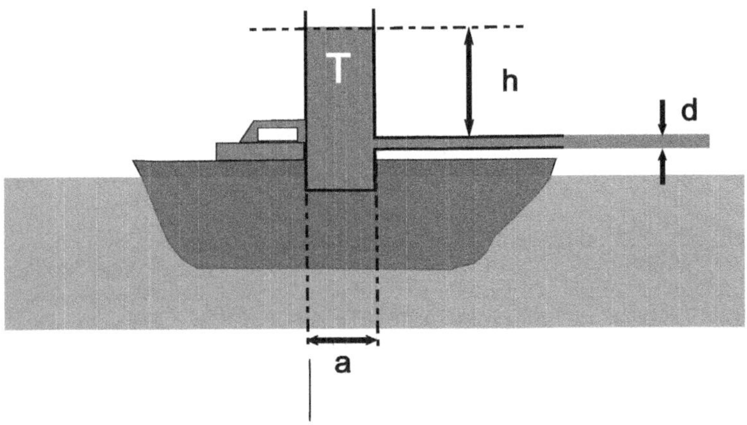

Figure 6.8. *A baby's bath toy*

The resistance to the forward motion of the boat in the bath is a force F directed along the axis of the boat and of intensity $F = kV$, where k is a constant and V is the velocity of the boat relative to the bath. T is filled with bath water up to a height of h above the level of the horizontal tube.

M is the mass of the boat including the mass of the water up to the height h.

The boat is stopped from moving forward.

1) Find the value of the thrust P exerted by the jet on the boat.

2) The boat is released at the initial moment

We assume that the expression found in (1) for the thrust P remains valid in this question when the boat is in movement. This implies that h varies very slowly. Moreover, we neglect the loss of mass in T due to the emptying of the boat through *the horizontal tube. This also involves assuming that the formula that gives the thrust remains valid in the framework of the boat, which has stopped being rigorously inertial.*

2.1) Determine the differential equation that governs the velocity of the boat.

2.2) Express this velocity as a function of time. Show that there is a maximum velocity.

2.3) Find the value of this maximum velocity.

2.4). Give the expression of the movement x (t) of the boat, which is equal to zero at the initial moment.

3) Orders of magnitude

We have:

$a = 10\,cm,\;\; b = 7\,cm,\;\; h = 20\,cm,\;\; d = 5\,mm,\;\; M = 1{,}9\,kg,\;\; m = 200\,g,$ density of the water $\rho = 1000\,kg.m^{-3}\;\; k = 0{,}8\,Nm^{-1}s$

3.1) Calculate the values of the maximum velocity found in (2) and (3).

3.2) Going back to the situation of question (2), calculate the time t_1 needed to reach 90% of the maximum velocity. Calculate the time $k = 0{,}8\,Nm^{-1}s$ needed to go from one end to the other of a 1.5 m bathtub at the maximum velocity found in (2).

From this, deduce an estimate of the amount of time needed for the boat to go from one end of the bathtub along its length.

Solution:

1) In this case, the framework of the boat is inertial. We can therefore calculate the exit velocity of the jet. This question has already been dealt with in the chapter on perfect fluids.

Considering the hypotheses, here we can reproduce a result obtained previously. The exit velocity V_{j1} of a horizontal jet resulting from a height of water is given by (the so-called Torricelli formula):

$$V_{j1} = \sqrt{2gh} \qquad [6.61]$$

The thrust is the result of a single "exiting" jet. It is given by:

$$P = q_m V_{j1} \qquad [6.62]$$

$$q_m = \rho s V_{j1}; \quad s = \pi \frac{d^2}{4} \qquad [6.63]$$

$$P = 2\rho g h \pi \frac{d^2}{4} \qquad [6.64]$$

2) Determining the movements

2.1) The forces applied to the boat are:

– the resistance force $\vec{F}$, of intensity $F = kV$;

– the thrust caused by the jet of liquid;

– Newton's second law applied to the boat, projected along an axis directed from the stern to the bow of the boat (therefore in the direction of the movement) is written as:

$$M\frac{d^2x}{dt^2} = P - F \qquad [6.65]$$

$$M\frac{d^2x}{dt^2} + k\frac{dx}{dt} = 2\rho g h \pi \frac{d^2}{4} \qquad [6.66]$$

It is always recommended to introduce several constants when dealing with this sort of equation to make the writing lighter.

We define $\alpha = \dfrac{k}{M}$ and $\beta = \dfrac{\rho g h \pi d^2}{2M}$ [6.67]

The equation for $x(t)$ becomes:

$$\frac{d^2x}{dt^2} + \alpha\frac{dx}{dt} = \beta \qquad [6.68]$$

with the initial conditions:

$$t = 0;\ x = 0;\ \frac{dx}{dt} = 0 \qquad [6.69]$$

Here we are looking at the velocity:

$$V(t) = \frac{dx}{dt} \quad [6.70].$$

The equation can be written again for $V(t)$. It is then a first-order equation and only has a single limit condition:

$$\frac{dV}{dt} + \alpha V = \beta \quad [6.71]$$

$$t = 0;\ V = 0 \quad [6.72]$$

2.2) The complete solution for [6.72] is the sum of the general equation without the second member $V_1(t)$ and of a particular solution to the complete equation $V_2(t)$. Resolution shows two constants that the conditions at the limits allow us to define.

Thus, the equation without the second member is:

$$\frac{dV_1}{dt} + \alpha V_1 = 0 \quad [6.73]$$

$$\frac{\frac{dV_1}{dt}}{V_1} = -\alpha \quad [6.74]$$

$$Log\, V_1 = -\alpha t + Log\ C_1 \quad [6.75]$$

$$V_1 = C_1 \exp - \alpha t \quad [6.76]$$

An expression of V_2 is obvious: $V_2 = \dfrac{\beta}{\alpha}$ [6.77]

The general expression of $V(t)$ is therefore:

$$V(t) = V_1 + V_2 = C_1 \exp - \alpha t + \frac{\beta}{\alpha} \quad [6.78]$$

By applying the condition at the limits:

$$t=0;\ V=0=C_1\exp-\alpha*0+\frac{\beta}{\alpha};\ C_1=-\frac{\beta}{\alpha} \qquad [6.79]$$

$$V(t)=\frac{\beta}{\alpha}(1-\exp-\alpha t) \qquad [6.80]$$

When $t \to \infty$, the exponential $\exp-\alpha t \to 0$, and the maximum velocity can be deduced:

$$\lim_{t \to \infty} V(t)=\frac{\beta}{\alpha} \qquad [6.81]$$

2.3) This maximum velocity is actually only reached after an infinite amount of time, meaning that it is nearly reached after a very long time. This maximum velocity appears when the acceleration of the boat becomes equal to zero. In this case, the two antagonist forces applied to the boat balance each other out. In other terms:

$$P=2\rho gh\pi\frac{d^2}{4}=kV;V=\frac{\rho gh\pi d^2}{2k} \qquad [6.82]$$

By introducing the two constants

$$\alpha=\frac{k}{M}\ \text{and}\ \beta=\frac{\rho gh\pi d^2}{2M} \qquad [6.83]$$

we immediately find:

$$V=\frac{\rho gh\pi d^2}{2M}*\frac{M}{k}=\frac{\beta}{\alpha} \qquad [6.84]$$

2.4) Going back to the complete equation for $x(t)$

$$\frac{d^2x}{dt^2}+\alpha\frac{dx}{dt}=\beta \qquad [6.85]$$

with the initial conditions:

$$t = 0; \;\; x = 0; \;\; \frac{dx}{dt} = 0 \qquad [6.86]$$

This equation is a classic in point mechanics. The solution is the sum of the general equation without the second member $x_1(t)$ and of a particular solution of the complete equation $x_2(t)$. Resolution shows two constants that the conditions at the limits will allow us to determine:

$$\frac{d^2 x_1}{dt^2} + \alpha \frac{dx_1}{dt} = 0 \qquad [6.87]$$

$$\frac{\frac{d^2 x_1}{dt^2}}{\frac{dx_1}{dt}} = -\alpha \qquad [6.88]$$

$$Log \frac{dx_1}{dt} = -\alpha t + Log\, C_1 \qquad [6.89]$$

$$\frac{dx_1}{dt} = C_1 \exp - \alpha t \qquad [6.90]$$

$$x_1 = -\frac{C_1}{\alpha} \exp - \alpha t + C_2 \qquad [6.91]$$

$$\frac{d^2 x_2}{dt^2} + \alpha \frac{dx_2}{dt} = \beta \qquad [6.92]$$

A particular solution for x_2 is obvious: $x_2 = \dfrac{\beta t}{\alpha}$ [6.93]

The general expression of $x(t)$ is therefore:

$$x(t) = x_1 + x_2 = -\frac{C_1}{\alpha} \exp - \alpha t + C_2 + \frac{\beta t}{\alpha} \qquad [6.94]$$

By applying the conditions at the limits:

$$t=0;\ -\frac{C_1}{\alpha}\exp-\alpha*0+C_2+\frac{\beta*0}{\alpha}=0\Rightarrow C_2=\frac{C_1}{\alpha} \qquad [6.95]$$

$$t=0;\ C_1\exp-\alpha*0+\frac{\beta}{\alpha}=0\Rightarrow\ C_1=-\frac{\beta}{\alpha};\ C_2=-\frac{\beta}{\alpha^2} \qquad [6.96]$$

The expression of $x\ (t)$ is:

$$x(t)=\frac{\beta}{\alpha^2}\exp-\alpha t-\frac{\beta}{\alpha^2}+\frac{\beta t}{\alpha}=\frac{\beta}{\alpha^2}\left[(1-\exp-\alpha t)+\alpha t\right] \qquad [6.97]$$

3) Orders of magnitude

3.1) The maximum velocity is:

$$\alpha=\frac{k}{M}=0,421;\ \beta=\frac{\rho gh\pi d^2}{2M}=4,055.10^{-2} \qquad [6.98]$$

$$V=\frac{\beta}{\alpha}=9,63.10^{-2}\ m.s^{-1}=9,63\,cm.s^{-1} \qquad [6.99]$$

3.2) To reach 90% of the maximum velocity, there must be:

$$V(t)=\frac{\beta}{\alpha}(1-\exp-\alpha t) \qquad [6.100]$$

$$0,9=1-\exp-\alpha t_1;\ \alpha t_1=-Log\,0,1=2,3 \qquad [6.101]$$

$$t_1=5,47\,s \qquad [6.102]$$

To cross the bathtub at the maximum limit, it takes:

$$t_2=\frac{1,5}{9,63.10^{-2}}=15,57\,s \qquad [6.103]$$

which is slightly longer than the time $t_1=5,47\,s$.

This means there is a time t_2 needed to cross the bathtub, which is in the order of 16–20 s.

6.4. Thrust exerted by a jet on a fixed wall

When a jet is directed on a wall, experiment shows that the wall tends to move away from the jet. The simple example of raindrops landing on a leaf is enough to illustrate this concept.

6.4.1. *Calculation of the thrust applied to a wall by a jet*

To calculate the value of the force thus applied by the jet, which is the thrust $\vec{P}$, we use the model of a panel held in place by a support. We note now that by applying Euler's theorem, we are not directly calculating the thrust $\vec{P}$ but rather the reaction $\vec{R}$ of this support, which is antagonist to the thrust so as to counterbalance the effect.

Therefore:

$$\vec{P} = -\vec{R} \,;\, P = R \qquad [6.104]$$

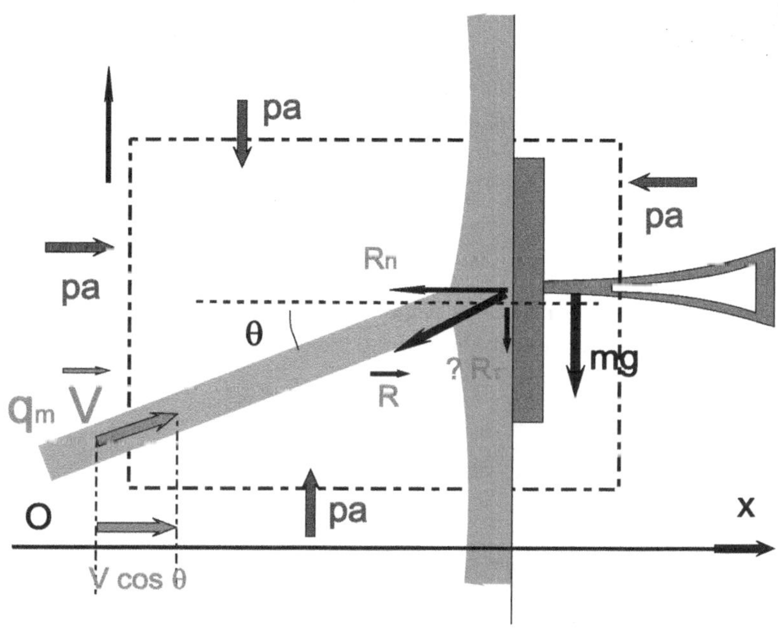

Figure 6.9. *Thrust of a jet on a wall at an oblique angle of incidence: we determine the reaction of the wall on the fluid*

The wall is not necessarily normal to the jet. Here we choose a jet whose angle with the normal to the panel is θ.

Furthermore, the kinematics of the jet can be modeled simply: here we choose a monokinetic jet, or in other words a uniform incident flow. However, the kinematics of the fluid flow leaving the panel is particularly complex, with poorly understood kinematics. Thus, as the exiting film is thin, we shall assume that the flow is "practically parallel" to the wall throughout.

It is therefore a good idea here to project Euler's theorem on an axis Ox perpendicular to the wall. This axis is orientated in the direction of the flow. To simplify the problem in terms of writing out the forces of gravity, we assume that the Ox axis is horizontal.

In this way, all of the velocity vectors of the exiting flow have a projection of zero and the "exiting quantity of movement flow" also has a projection of zero along this axis Ox.

We choose a reference surface that includes the panel, which is a parallelepiped whose faces upstream and downstream are perpendicular to the axis Ox.

In this way, the reaction of the support on the panel is integrated with the volume forces applied to "everything that is in the reference surface".

The incident flow is easily calculable. S is the section of the incident jet:

$$q_m = \rho SV \qquad [6.105]$$

NOTE: The velocity vector makes an angle θ with the panel, and therefore with the axis Ox. As a result, the contribution of the incident flow to the projection on Ox of the integral:

$$\iint_S \vec{V}\, dq_m \text{ is } \mathrm{Proj}_{Ox}\left[\iint_S \vec{V}\, d\,q_m\right] = -q_m\left(V\cos\theta\right) \qquad [6.106]$$

The "–" sign comes from the fact that the unit vectors normal to the elementary surfaces dS of the reference surface crossed by the incident jet are directed in the direction opposite to that of the axis Ox.

Following the previous reasoning, all of the contributions of the exiting flow at $\mathrm{Proj}_{Ox}\left[\iint_S \vec{V}\, d\, q_m\right]$ are counted as positive.

In terms of the surface forces, they are limited to the forces of pressure on the upstream face and the downstream face. They balance each other out mutually.

As the other faces are parallel to $O\,x$, the pressure forces have a component of zero along this axis.

As the axis $O\,x$ is horizontal, the forces of gravity have a projection equal to zero on this axis.

The only force that contributes to the writing of the second member of the projected Euler's theorem on $O\,x$ is the reaction force of the axis. Here, we therefore only have access to the normal component R_n of this reaction, and as a result to the normal component of the panel of the thrust provided by the jet against the panel opposed to it, and therefore equal to the modulus: $P_n = R_n$:

$$\mathrm{Proj}_{Ox}\left[\iint_S \vec{F}_S\, dS\right] = 0 \qquad [6.107]$$

$$\mathrm{Proj}_{Ox}\left[\iiint_D \vec{F}_V\, d\omega\right] = -R_n \qquad [6.108]$$

Finally, we obtain the projection of Euler's theorem on $O\,x$:

$$\iint_S \vec{V}\, d\, q_m = \iiint_D \vec{F}_V\, d\omega + \iint_S \vec{F}_S\, dS \qquad [6.109]$$

$$-q_m\left(V\cos\theta\right) = -R_n = -P_n\, D \qquad [6.110]$$

$$P_n = \rho S V^2 \cos\theta \qquad [6.111]$$

For a normal jet, $\cos\theta = 1$ and:

$$P_n = \rho S V^2 \qquad [6.112]$$

It is important to note that for an oblique jet $\cos\theta \neq 1$ the cosine is not squared, unlike what an overzealous extension of the previous formula to the case of an oblique jet might suggest.

The normal jet is the only case where the thrust is fully known. In the case of the oblique jet, the flow exiting parallel to the wall is asymmetric. This results in a projection of the flow of the quantity of movement along any axis perpendicular to Ox that cannot be equal to zero. The complexity of the calculation means we will steer clear of this question.

6.4.2. *Jet impacting on a wall*

EXAMPLE 6.5 (A TV game).–

During one of those popular games shown on television, various European cities are playing against each other. In the current game we are looking at, it is a French team versus a Belgian team.

Principle of the game. A cart C can roll without friction along horizontal rails. A vertical panel P is fixed to the cart. Thus equipped, the whole system has a mass m. Each of the two teams is placed on different sides of the panel. Each team is given a fire hose. The hoses of the two teams are linked to a single reservoir, whose internal pressure is maintained at a value of p_R, which is obviously higher than the atmospheric pressure p_a. The water is considered here to be a perfect fluid.

Each team directs its jet normally onto different sides of the panel.

The goal of the game, which may be obvious at this stage, is to push the cart into the opponent's area.

To make things more interesting, each team has to choose the size of the diameter of the nozzle of their hose orifice at the beginning of the game.

Two diameters are available: D and $D\sqrt{2}$.

The French team chooses the diameter D and the Belgian team chooses the diameter $D\sqrt{2}$.

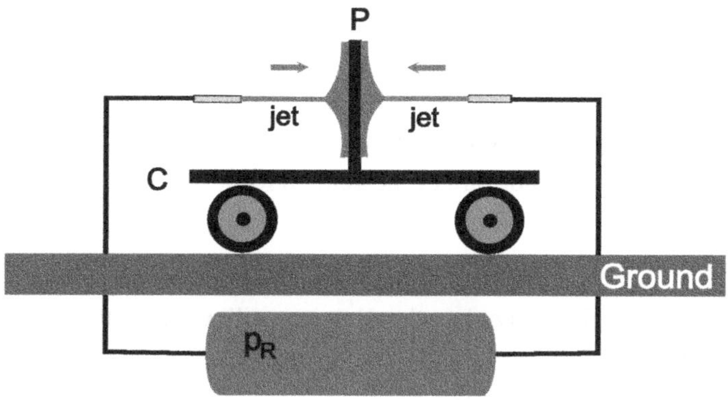

Figure 6.10. *A TV game*

1) Each team directs its jet onto its own side of the panel. Which team should win in theory?

2) The winning team then lowers its jet by an angle of θ with the horizontal. What must this angle be for there to be a draw?

3) The two teams put their jets back against the panel normally. The cart, which was initially immobile, starts a movement of variable velocity $u(t)$ orientated toward the losers. The positions are located in terms of an axis Ox from left to right, assuming the losers are on the right.

Give the differential equation that allows us to calculate $u(t)$. For this, we will need to use a nearly stationary procedure for the calculation of the thrust. Why is this?

4) Is there *a priori* a limit value for $u(t)$?

5) Provide the expression of $u(t)$. Find the result from question (4).

Solution

1) The velocity V of the incident jet on each side of the panel can be calculated from the Bernoulli theorem. The calculation is simple. Ignoring the differences in altitude, there is immediately the same velocity of the jet on each side of the panel. A classic result is obtained here:

$$p_R = p_a + \rho\frac{V^2}{2} \qquad [6.113]$$

$$V = \sqrt{2\frac{p_R - p_a}{\rho}} \quad [6.114]$$

The corresponding thrusts on either side of the panel P_F and P_B, respectively, on the French and Belgian side, are of the general form:

$$P = \rho S V^2 \quad [6.115]$$

We can see straight away that the Belgian will win, as they have a greater surface of the jet.

There is:

For the French:

$$P_F = \rho S_F V^2 \quad [6.116]$$

$$S_F = \pi \frac{D^2}{4} \quad [6.117]$$

$$P_F = \rho\ 2\left(\frac{p_R - p_a}{\rho}\right)\pi\frac{D^2}{4} \quad [6.118]$$

For the Belgians:

$$P_F = \rho S_B V^2 \quad [6.119]$$

$$S_B = \pi \frac{D^2}{2} \quad [6.120]$$

$$P_B = \rho\ 2\left(\frac{p_R - p_a}{\rho}\right)\pi\frac{D^2}{2} \quad [6.121]$$

2) The Belgians incline their jet by an angle θ. Their thrust becomes:

$$P'_B = \rho\ 2\left(\frac{p_R - p_a}{\rho}\right)\pi\frac{D^2}{2}\cos\theta \quad [6.122]$$

P'_B becomes equal to P_F for

$$\frac{\cos\theta}{2}=\frac{1}{4};\ \cos\theta=0,5;\ \theta=\frac{\pi}{4} \qquad [6.123]$$

The Belgians must incline their jet by 45°

3) The form used in the calculation of the thrust assumes that the panel is immobile, and therefore that the reference frame attached to it is Galilean. For small variations, we can maintain this formula for the current question, but it will only serve as an approximation.

The relative velocities of the jets in relation to the panel become:

on the Belgian side: $V-u$

on the French side: $V+u$

As the velocity keeps the same value previously calculated:

$$V=\sqrt{2\frac{p_R-p_a}{\rho}} \qquad [6.124]$$

The thrusts then become

on the Belgian side:

$$\rho S_B (V-u)^2 \qquad [6.125]$$

on the French side:

$$\rho S_F (V+u)^2 \qquad [6.126]$$

The resulting force F is the difference between the two.

By noting that $S_B = 2S_F = 2\pi\frac{D^2}{4}$ [6.127]

$$F=\rho S_B (V-u)^2 - \rho S_F (V+u)^2 = \rho S_F \left[2(V-u)^2 - (V+u)^2\right] \qquad [6.128]$$

The fundamental equation of dynamics gives us the differential equation that satisfies $u(t)$:

$$\frac{du}{dt} = F \Rightarrow m\frac{du}{dt} = \rho S_F\left[2(V-u)^2 - (V+u)^2\right] \quad [6.129]$$

associated with the clear condition at the limits:

$$t = 0;\ u = 0 \quad [6.130]$$

The expression of F can be rearranged:

$$F = \rho S_F\left[2(V-u)^2 - (V+u)^2\right] = \rho S_F\left[2V^2 + 2u^2 - 4Vu - V^2 - u^2 - 2Vu\right] \quad [6.131]$$

$$F = \rho S_F\left[V^2 + u^2 - 6Vu\right] \quad [6.132]$$

4) Starting from a value of zero, $u(t)$ increases under the effect of the force $\vec{F}$. This force is proportional to the difference of the two terms, $(V-u)^2 - (V+u)^2$ the first decreasing and the second increasing. $\vec{F}$ is therefore decreasing. When it is equal to zero, u reaches a limit value.

$\vec{F}$ cancels itself out for two values of u, roots of $(V-u)^2 - (V+u)^2 = V^2 + u^2 - 6Vu$. Using the classical resolution of algebraic second-degree equations, these two roots come out as:

$$u_1 = \left(3 - 2\sqrt{2}\right)V \quad [6.133.a]$$

and $$u_2 = \left(3 + 2\sqrt{2}\right)V \quad [6.133.b]$$

It is clear that the limit velocity is the smallest of these values reached before the other as $u(t)$ increases. Therefore, the limit velocity is:

$$u_{\lim} = u_1 = \left(3 - 2\sqrt{2}\right)V \quad [6.134]$$

The following shall be useful later:

$$(V-u)^2 - (V+u)^2 = V^2 + u^2 - 6Vu = (u-u_1)(u-u_2) \quad [6.135]$$

5) The equation can be written as follows:

$$\frac{du}{dt} = \frac{\rho S_F}{m}[V^2 + u^2 - 6Vu] \quad [6.136.a]$$

$$\frac{du}{u^2 - 6Vu + V^2} = \alpha\, dt \quad [6.136.b]$$

$$\alpha = \frac{\rho S_F}{m} \quad [6.137]$$

Resolution implies the decomposition of:

$$\frac{1}{u^2 - 6Vu + V^2} = \frac{1}{(u-u_1)(u-u_2)} \quad [6.138.a]$$

into simple elements

$$\frac{1}{u^2 - 6Vu + V^2} = \frac{A}{(u-u_1)} + \frac{B}{(u-u_2)} \quad [6.138.b]$$

There is:

$$\frac{1}{u^2 - 6Vu + V^2} = \frac{A}{(u-u_1)} + \frac{B}{(u-u_2)} = \frac{1}{u_1 - u_2}\left[\frac{1}{(u-u_1)} - \frac{1}{(u-u_2)}\right] \quad [6.139]$$

The differential equation is written as:

$$\frac{du}{u_1 - u_2}\left[\frac{1}{(u-u_1)} - \frac{1}{(u-u_2)}\right] = \alpha\, dt \quad [6.140]$$

$$\frac{du}{(u-u_1)} - \frac{du}{(u-u_2)} = \beta\, dt \quad [6.141]$$

$$\beta = \alpha\,(u_1 - u_2) \quad [6.142]$$

By integrating the two terms and taking into account the initial condition:

$$Ln(u - u_1) - Ln(u - u_2) = \beta t + Ln\ C \qquad [6.143]$$

$$\frac{u - u_1}{u - u_2} = C \exp \beta t \qquad [6.144]$$

$$t = 0;\ u = 0 \Rightarrow C = \frac{u_1}{u_2} \qquad [6.145]$$

For a large value of time, $t \to \infty;\ \dfrac{u - u_1}{u - u_2} \to \infty$ [6.146]

Therefore: $u - u_2 \to 0; u \to u_2$ [6.147]

And this provides the value given in [6.134]

$$u_{\lim} = u_2 = \left(3 - 2\sqrt{2}\right) V \qquad [6.148]$$

It is important to note that this limit velocity is theoretically only reached after an infinite amount of time. In practice, it cannot be reached, since the players of the two teams are necessarily at a finite distance in front and behind the moving cart.

6.5. Other applications for Euler's theorems

6.5.1. *Application of Euler's theorem to a head loss calculation*

EXAMPLE 6.6 (Head loss in a sudden enlargement).–

A loss of charge is a loss of energy. In this theory, attributed to Borda and Carnot, we can see that the principle loss of energy is not due to viscosity.

Two horizontal cylindrical tubes T_1 and T_2 are joined suddenly. Their sections are S_1 upstream and S_2 downstream, respectively. The ratio $\sigma = \dfrac{S_1}{S_2}$ is given with $\sigma < 1$.

We admit the following hypotheses:

The fluid stream has the geometry shown in the figure; it should be noted that it sticks back to the wall after a certain distance from the singularity. Outside of this stream, there is a "dead fluid", the kinematics of which shall be described later.

The flow is uniform, of velocity V_1 upstream of the section change and becomes uniform again, with a velocity of V_2 as soon as it sticks again downstream of the singularity. The volume forces, including the forces of gravity, have a negligible effect.

The static pressure in the entrance plane of the second tube (meaning the plane of the joint of the two tubes) is constant and equal to the pressure present in a section of the tube T_1 located immediately upstream of the singularity; p_1 is this pressure and p_2 is the pressure present in the section S_2 after it is stuck again.

We assume the fluid is perfect.

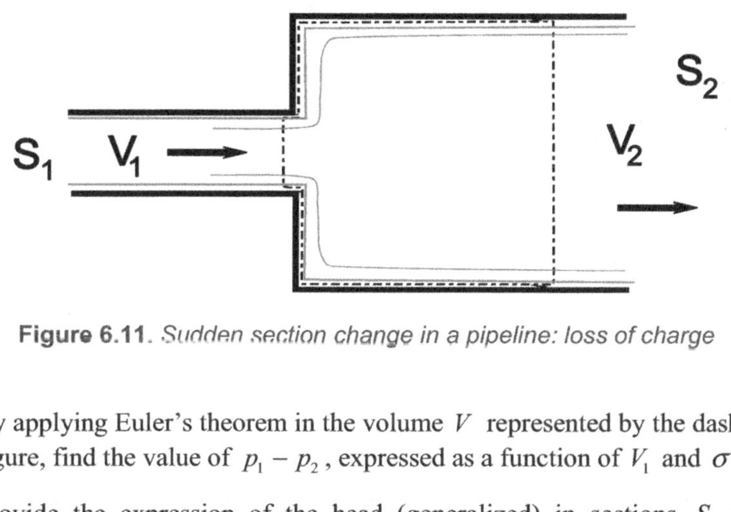

Figure 6.11. *Sudden section change in a pipeline: loss of charge*

1) By applying Euler's theorem in the volume V represented by the dashed lines in the figure, find the value of $p_1 - p_2$, expressed as a function of V_1 and σ.

2) Provide the expression of the head (generalized) in sections S_1 and S_2. Deduce the expression of the singular loss of charge ΔH_S produced by the section change, as well as the expression of the singular head loss coefficient ζ. What does this formula remind you of?

3) Question without calculations. We shall now look at the worth of this formula for real fluids.

3.1) Assume again that the fluid is perfect. Draw the form of the current lines of the flow in the "dead fluid" zone. Is there not something rather surprising about this drawing?

3.2) Assume now that the fluid is real. This would suggest that the "dead fluid" is actually being recirculated. Despite this, we shall assume that the transversal pressure gradients remain negligible: why is it acceptable to do this?

In this context of a hypothesis, why can we keep the formula established in (2)?

Where is the loss of energy localized in the two hypotheses of a perfect fluid and of a real fluid?

Solution

1) Ox is the common axis of the two tubes, oriented from upstream to downstream, so from left to right. For a perfect fluid, the only forces that must be taken into account are the forces of pressure whose summary projected onto Ox is $\left(p_1\, S_2 - p_2\, S_2\right) = S_2\left(p_1 - p_2\right)$ and the forces of gravity whose horizontal projection is equal to zero.

The flow is uniform on each section and the mass flow is equal to $q_m = \rho\, S_1\, V_1 = \rho\, S_2\, V_2$, entering flow for S_1 and exiting for S_2. The summary of the "exiting" quantity of movement is $q_m\left(V_2 - V_1\right)$.

Let us project Euler's first theorem on the axis of the tubes system:

$$\iint_S \vec{V}\, dq_m = \iiint_D \vec{F}_V\, d\omega + \iint_S \vec{F}_S\, dS \qquad [6.149]$$

$$q_m\left(V_2 - V_1\right) = S_2\left(p_1 - p_2\right) \qquad [6.150]$$

$$p_1 - p_2 = \frac{q_m\left(V_2 - V_1\right)}{S_2} \qquad [6.151]$$

2) By definition, the generalized charges H_1 and H_2 in the sections S_1 and S_2 are written as:

$$H_1 = \frac{V_1^2}{2g} + \frac{p_{G1}}{\varpi} \qquad [6.152]$$

$$H_2 = \frac{V_2^2}{2g} + \frac{p_{G2}}{\varpi} \qquad [6.153]$$

with $p_{G1} = p_1 + \rho g Z;\ p_{G2} = p_2 + \rho g Z\ \varpi = \rho g$ [6.154]

Z is the applicate of the points. Here the problem is simplified by assuming that the tubes are moderated sections and that the changes in the cotes play a very small role in the summary of the forces.

The difference between H_1 and H_2 is the loss of head ΔH_S between the sections S_1 and S_2:

$$\Delta H_S = H_1 - H_2 = \left(\frac{V_1^2}{2g} + \frac{p_{G1}}{\varpi}\right) - \left(\frac{V_2^2}{2g} + \frac{p_{G2}}{\varpi}\right) \quad [6.155]$$

The preservation of the mass flow allows us to write:

$$q_m = \rho S_1 V_1 = \rho S_2 V_2 \quad [6.156]$$

$$\frac{V_2}{V_1} = \frac{S_1}{S_2} = \sigma \quad [6.157]$$

Some simple operations allow us to find ΔH_S and ζ:

$$\Delta H_S = H_1 - H_2 = \left(\frac{V_1^2}{2g} + \frac{p_{G1}}{\varpi}\right) - \left(\frac{V_2^2}{2g} + \frac{p_{G2}}{\varpi}\right)$$

$$= \frac{V_1^2 - V_2^2}{2g} + \frac{(p_1 + \rho g Z) - (p_2 + \rho g Z)}{\rho g} \quad [6.158]$$

$$\Delta H_S = \frac{V_1^2 - V_2^2}{2g} + \frac{p_1 - p_2}{\rho g} = \frac{V_1^2 - V_2^2}{2g} + \frac{q_m (V_2 - V_1)}{\rho g S_2} = \frac{V_1^2 - V_2^2}{2g} + \frac{\rho S_2 V_2 (V_2 - V_1)}{\rho g S_2} \quad [6.159]$$

$$\Delta H_S = \frac{1}{2g}\left[\left(V_1^2 - V_2^2\right) + 2V_2\left(V_2 - V_1\right)\right] \quad [6.160]$$

$$\Delta H_S = \frac{1}{2g}\left(V_1^2 + V_2^2 - 2V_1V_2\right) = \frac{\left(V_1 - V_2\right)^2}{2g} = \frac{V_1^2}{2g}\left(1 - \frac{V_2}{V_1}\right)^2 = \frac{V_1^2}{2g}(1-\sigma)^2 \quad [6.161]$$

$$\Delta H_S - \varsigma \frac{V_1^2}{2g} \quad [6.162]$$

The formula to calculate ς, which appears in a large number of hydraulics treaties, is:

$$\varsigma = (1 - \sigma)^2 \qquad [6.163]$$

3) In this question, the kinematic model of a perfect fluid is opposed to the kinematic model observed generally in a real fluid.

3.1) The kinematic model in a "perfect fluid" assumes that the current lines follow the walls (the normal components of the velocity for fluids in contact with this wall are equal to zero). This results in a "right angle" flow, which can seem strange in terms of the physics.

In this model of the loss of charge, which we must remind ourselves is a loss of energy, is located the loss of kinetic energy at the section change. This is also what happens in orifice systems such as nozzles or diaphragms.

3.2) In the model of the perfect fluid, we consider more realistic kinematics. The fluid stream has a "conical" form and moves progressively from one section to another. The dead fluid zone is in recirculation.

We can go back to the calculation carried out in the two previous questions, and notably the application of Euler's theorem, to find and accept the same result except this time with a more satisfying approximation.

In this recirculation, we consider that the streamlines of the flow are practically lines parallel to the axis of the tube. In this case, the generating pressure p_G is preserved over any section of the system that is perpendicular to the axis of the tubes, and the calculation can be carried out identically. The fact that the direction of the flow is inversed on some of these lines in relation to others, due to the phenomenon of recirculation, does not change anything.

Formula [6.163] can therefore be applied in the case of perfect fluids. It is of practical use.

It could be surprising to see the term loss of charge being used for a calculation carried out in a perfect fluid. There is a loss of energy that is not caused by a phenomenon of viscosity. This loss happens due to the "friction" of a fluid that makes a right angle turn at the section change. For a real fluid, the recirculation implies in the same way a loss of energy due to the section change. However, the viscosity then consumes the mechanical energy of the fluid that is recirculating.

6.5.2. *A case for the application of Euler's second theorem*

To illustrate the advantages of Euler's second equation, here we propose a simple exercise where it can be applied.

We propose using two strategies for resolution in dealing with this problem: one that relies on Euler's first theorem and the other on the second.

EXAMPLE 6.7 (Modeling of a garden sprinkler).–

A garden sprinkler is made of an assembly of two tubes T of diameter $d = 7{,}5\,mm$ supplied with water by a pump. This pump delivers a flow of $q_V = 2\,400\ litres.hr^{-1}$, which is shared in two equal parts between the two tubes *T*. The density of the water is $\rho = 1000\,kg.m^{-3}$.

The two tubes can turn in the horizontal plane. Each tube is made up of two perpendicular parts. The longest, T_1, of radius rayon $R = 75\,cm$ passes through the rotation center of the system, while the shortest, T_2, is therefore a tangent to the circular trajectory of the bend of T. The two branches T_1 are located on the same line, which therefore constitutes the diameter of the trajectory of the bends. They meet in the point O, which is the rotation center of the system. The extremities of the two branches T_1 are located on either side of this diameter. Through the two jets flowing in the tubes T_2, a reaction system determines a rotational movement of the sprinkler.

All of the friction of such a system can be seen in the form of resistant couple of constant $C = 1{,}8\,N.m.Rd^{-1}.s$. The rotational velocity of the sprinkler (ω)is observed to be constant.

Later, a vertical axis directed upward will be defined, whose origin is O.

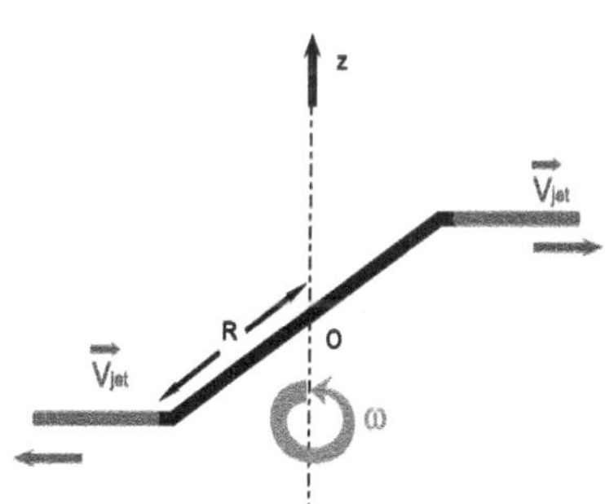

Figure 6.12. *A simplified garden sprinkler*

1) Resolution using Euler's first theorem

1.1) Determining the thrust

1.1.1) Assuming the fluid is perfect, what water pressure is needed to supply the sprinkler?

1.1.2) What is the thrust provided for each tube T_2?

1.2) Noting that the rotational velocity is constant, calculate ω. Express this velocity in rotations per minute.

2) Resolution using Euler's second theorem

2.1) Calculate the flow of the "moment of the quantity of movement" projected onto Oz.

2.2) Deduce the rotational velocity ω.

Solution

1) Application of Euler's first theorem

1.1.1) This question calls for the calculation of the exit velocity of each jet.

Each jet has a flow equal to half of q_V. The ejection velocity V_{jet} is therefore:

$$V_{jet} = \frac{q_V}{2\frac{\pi d^2}{4}} = \frac{2q_V}{\pi d^2} \qquad [6.164]$$

$$q_V = 2400\, litres.hr^{-1} = 6,67.10^{-4}\, m^3.s^{-1} \quad ; \quad \frac{\pi d^2}{4} = 4,42.10^{-5}\, m^2 \qquad [6.165]$$

$$V_{jet} = 7,55\, m.s^{-1} \qquad [6.166]$$

By application of the Bernoulli theorem, written in the framework of tube T_2, this velocity means the pressure of the network is such that:

$$p_R - p_a = \frac{\rho V_{jet}^2}{2} \qquad [6.167]$$

With:

$$p_R = \frac{\rho V_{jet}^2}{2} + p_a = 1,281.10^5\ Pa = 1,3 bar \quad [6.168]$$

1.1.2) The thrust resulting from each jet from each tube T_2 is calculated from Euler's first theory, as shown in sections 6.2.1 and 6.2.3.

The thrust of each jet is:

$$P_{jet} = q_m V_{jet} = \rho \frac{q_V}{2} V_{jet} \quad [6.169]$$

$$P_{jet} = 1000 * \frac{6,67.10^{-4}}{2} * 7,55 = 2,52\ N \quad [6.170]$$

This thrust is the same for each of the two jets.

We must note that implicitly through this formula we are applying Euler's first theorem in the reference frame of T_2, where the flow is stationary. Rigorously speaking, this framework is not inertial. Volume forces (forces of inertia) must be considered. We would recommend going back to the proof of 6.2.1 and 6.2.3. It can be noted that the forces of inertia are directed along the radius of the trajectory circle, and their component on the axis where the fundamental principle of dynamics is projected is equal to zero. The visible effect of these inertia forces is to curve the jet toward the outside downstream of its exit point. This does not affect the calculation of the thrusts responsible for rotation.

1.2) The rotational velocity is constant, so therefore the forces of reaction and of friction must balance out, which we shall write while remaining in the horizontal plane of rotation. The equilibrium to be written is reduced down to these forces, as the antagonist forces of pressure in the system whose support is on the plane of rotation compensate each other. Furthermore, the forces of gravity are perpendicular to the plane of rotation.

The thrust forces have a resulting force that is equal to zero as the thrusts of the two jets balance each other. The equilibrium must be written in terms of moments, as there is a moment that is the result of friction. Let us look at the moments in relation to O of the two forces of thrust and friction. They are represented by three vectors carried on Oz. The two thrusts are directed upward and the frictions are directed downward.

The whole must cancel itself out, and in terms of norms, we write:

$$2P_{jet}R - C\omega = 0 \quad [6.171]$$

which gives the angular velocity of rotation:

$$\omega = \frac{2P_{jet}R}{C} = \frac{2R\rho q_V V_{jet}}{2C} \quad [6.172.a]$$

$$\omega = \frac{2P_{jet}R}{C} = \frac{2*2,52*0,75}{1,8} = 2,1\,Rd.s^{-1} \quad [6.172.b]$$

which gives the value:

$$\omega = \frac{2,1}{2\pi}*60 = 20\,rotations.mn^{-1} \quad [6.173]$$

2) Application of Euler's second theorem

2.1) The flow of the "moment of the quantity of movement" is none other than the moment of the quantity of movement of the mass of the fluid ejected by the tubes T_2.

We consider the quantity of the movement of the fluid ejected per second from a T_2. The mass ejected per second is q_m, it has a velocity vector $\vec{V}_{jet}$, located in the horizontal plane containing the origin and the normal to the vector position $\vec{r}$ of the bend of T. It is important to note that here the norm of $\vec{r}$ is R.

The flow of the "moment of the quantity of movement" relative to each T_2 is therefore $q_m\vec{V}_{jet}$ and its moment in relation to O is a vector:

$$\vec{M}_{jet} = \vec{r} \wedge q_m \vec{V}_{jet} \quad [6.174]$$

To take both jets into account, the flow of the "moment of the quantity of movement" is equal to twice this vector.

This "flow vector" is carried by Oz. Its norm is:

$$2M_{jet} = 2R\,q_m\vec{V}_{jet} \quad [6.175]$$

2.2) The moment of the forces of friction is also carried by Oz, and its norm is:

$$M_f = C\omega \qquad [6.176]$$

Euler's theorem is therefore written as, projected onto Oz:

$$2M_{jet} = 2R\,q_m\,V_{jet} = C\,\omega \qquad [6.177]$$

Hence, the value of ω, already found in 1), is:

$$\omega = \frac{2R\,q_m\,V_{jet}}{C} \qquad [6.178]$$

Bibliography

[AND 16] ANDERSON J., *Fundamentals of Aerodynamics*, McGraw Hill, 2016.

[ASP 74] ASPNES D.E., "Optimizing precision of rotating-analyzer ellipsometers", *Journal of the Optical Society of America*, vol. 64, pp. 639–646, 1974.

[BAT 12] BATCHELOR G.K., *An Introduction to Fluid Dynamics*, Cambridge University Press, 2012.

[BIR 60] BIRD R.B., STEWART W.E., LIGHTFOOT E.N., *Transport Phenomena*, John Wiley & Sons, 1960.

[BRU 68] BRUN E.A., MATTHIEU J.-P., MARTINOT-LAGARDE A., *Mécanique des fluides*, vols 1–3, Dunod, Paris, 1968.

[BRU 08] BRUN J.M., *Théorie de la couche limite*, Gauthier Villars, Paris, 2008.

[CAN 02] CANDEL S., *Mécanique des fluides*, Dunod, Paris, 2002.

[COM 97] COMOLET R., *Mécanique expérimentale des fluides*, vols 1–3, Elsevier Masson, Paris, 1997.

[HIN 75] HINZE J.O., *Turbulence*, McGraw Hill, 1975.

[HOL 09] HOLMAN J.P., *Heat Transfer*, McGraw Hill, 2009.

[KNU 58] KNUDSEN J.G., KATZ D.L., *Fluid Dynamics and Heat Transfer*, McGraw Hill, 1958.

[LAN 59] LANDAU L., LIFCHITZ E., *Fluid Mechanics*, Pergamon Press, 1959.

[LAN 94] LANDAU L., LIFCHITZ E., *Mécanique des fluides*, Ellipses, Paris, 1994.

[LEO 85] LEONTIEV A., *Théorie des echanges de chaleur et de masse*, Mir, Moscow, 1985.

[NEK 67] NEKRASSOV B.B., *Cours d'hydraulique*, Mir, Moscow, 1967.

[NEK 68] NEKRASSOV B.B., *Hydraulics for Aeronautical Engineers,* Mir, Moscow, 1968.

[NEW 05] NEWTON I., *Philosophiæ Naturalis Principia Mathematica*, Dunod, Paris, 2005.

[RHY 09] RHYMING I.L., *Dynamique des fluides*, Presses Polytechniques Romandes, 2009.

[SCH 09] SCHLICHTING H., *Boundary Layer Theory*, Springer-Verlag, 2009.

[TEN 72] TENNEEKES H., LUMLEY J., *A First Course in Turbulence*, MIT Press, 1972.

Index

A, B, C

D, E, F

J, K, L

M, N, P

Q, R

S, T, V

Other titles from

iSTE

in

Mechanical Engineering and Solid Mechanics

2016

BOREL Michel, VÉNIZÉLOS Georges
Movement Equations 1: Location, Kinematics and Kinetics
(Non-deformable Solid Mechanics Set – Volume 1)

Movement Equations 2: Mathematical and Methodological Supplements
(Non-deformable Solid Mechanics Set – Volume 2)

BOYARD Nicolas
Heat Transfer in Polymer Composite Materials

CARDON Alain, ITMI Mhamed
New Autonomous Systems
(Reliability of Multiphysical Systems Set – Volume 1)

DAHOO Pierre Richard, POUGNET Philippe, EL HAMI Abdelkhalak
Nanometer-scale Defect Detection Using Polarized Light
(Reliability of Multiphysical Systems Set – Volume 2)

DE SAXCÉ Géry, VALLÉE Claude
Galilean Mechanics and Thermodynamics of Continua

DORMIEUX Luc, KONDO Djimédo
Micromechanics of Fracture and Damage (Micromechanics Set – Volume 1)

EL HAMI Abdelkhalak, RADI Bouchaib
Stochastic Dynamics of Structures (Mathematical and Mechanical Engineering Set – Volume 2)

GOURIVEAU Rafael, MEDJAHER Kamal, ZERHOUNI Noureddine
From Prognostics and Health Systems Management to Predictive Maintenance 1: Monitoring and Prognostics (Reliability of Multiphysical Systems Set – Volume 4)

KHARMANDA Ghias, EL HAMI Abdelkhalak
Reliability in Biomechanics (Reliability of Multiphysical Systems Set – Volume 3)

Biomechanics: Optimization, Uncertainties and Reliability (Reliability of Multiphysical Systems Set – Volume 5)

MOLIMARD Jérôme
Experimental Mechanics of Solids and Structures

RADI Bouchaib, EL HAMI Abdelkhalak
Material Forming Processes: Simulation, Drawing, Hydroforming and Additive Manufacturing (Mathematical and Mechanical Engineering Set – Volume 1)

2015

KARLIČIĆ Danilo, MURMU Tony, ADHIKARI Sondipon, MCCARTHY Michael
Non-local Structural Mechanics

SAB Karam, LEBÉE Arthur
Homogenization of Heterogeneous Thin and Thick Plates

2014

ATANACKOVIC M. Teodor, PILIPOVIC Stevan, STANKOVIC Bogoljub, ZORICA Dusan
Fractional Calculus with Applications in Mechanics: Vibrations and Diffusion Processes

ATANACKOVIC M. Teodor, PILIPOVIC Stevan, STANKOVIC Bogoljub, ZORICA Dusan
Fractional Calculus with Applications in Mechanics: Wave Propagation, Impact and Variational Principles

CIBLAC Thierry, MOREL Jean-Claude
Sustainable Masonry: Stability and Behavior of Structures

ILANKO Sinniah, MONTERRUBIO Luis E., MOCHIDA Yusuke
The Rayleigh–Ritz Method for Structural Analysis

LALANNE Christian
Mechanical Vibration and Shock Analysis – 5-volume series – 3rd edition
Sinusoidal Vibration – volume 1
Mechanical Shock – volume 2
Random Vibration – volume 3
Fatigue Damage – volume 4
Specification Development – volume 5

LEMAIRE Maurice
Uncertainty and Mechanics

2013

ADHIKARI Sondipon
Structural Dynamic Analysis with Generalized Damping Models: Analysis

ADHIKARI Sondipon
Structural Dynamic Analysis with Generalized Damping Models: Identification

BAILLY Patrice
Materials and Structures under Shock and Impact

BASTIEN Jérôme, BERNARDIN Frédéric, LAMARQUE Claude-Henri
Non-smooth Deterministic or Stochastic Discrete Dynamical Systems: Applications to Models with Friction or Impact

EL HAMI Abdelkhalak, BOUCHAIB Radi
Uncertainty and Optimization in Structural Mechanics

KIRILLOV Oleg N., PELINOVSKY Dmitry E.
Nonlinear Physical Systems: Spectral Analysis, Stability and Bifurcations

LUONGO Angelo, ZULLI Daniele
Mathematical Models of Beams and Cables

SALENÇON Jean
Yield Design

2012

DAVIM J. Paulo
Mechanical Engineering Education

DUPEUX Michel, BRACCINI Muriel
Mechanics of Solid Interfaces

ELISHAKOFF Isaac *et al.*
Carbon Nanotubes and Nanosensors: Vibration, Buckling and Ballistic Impact

GRÉDIAC Michel, HILD François
Full-Field Measurements and Identification in Solid Mechanics

GROUS Ammar
Fracture Mechanics – 3-volume series
Analysis of Reliability and Quality Control – volume 1
Applied Reliability – volume 2
Applied Quality Control – volume 3

RECHO Naman
Fracture Mechanics and Crack Growth

2011

KRYSINSKI Tomasz, MALBURET François
Mechanical Instability

SOUSTELLE Michel
An Introduction to Chemical Kinetics

2010

BREITKOPF Piotr, FILOMENO COELHO Rajan
Multidisciplinary Design Optimization in Computational Mechanics

DAVIM J. Paulo
Biotribolgy

PAULTRE Patrick
Dynamics of Structures

SOUSTELLE Michel
Handbook of Heterogenous Kinetics

2009

BERLIOZ Alain, TROMPETTE Philippe
Solid Mechanics using the Finite Element Method

LEMAIRE Maurice
Structural Reliability

2007

GIRARD Alain, ROY Nicolas
Structural Dynamics in Industry

GUINEBRETIÈRE René
X-ray Diffraction by Polycrystalline Materials

KRYSINSKI Tomasz, MALBURET François
Mechanical Vibrations

KUNDU Tribikram
Advanced Ultrasonic Methods for Material and Structure Inspection

SIH George C. *et al.*
Particle and Continuum Aspects of Mesomechanics

Printed and bound by CPI Group (UK) Ltd, Croydon, CR0 4YY

13/09/2026

14961021-0006